Steamed Breads

Related Titles

Genetic and Genomic Resources for Grain Cereals Improvement
(ISBN 978-0-12-802000-5)

Encyclopedia of Food Grains, 2nd Edition
(ISBN 978-0-12-394437-5)

Cereal Grains for the Food and Beverage Industries
(ISBN 978-0-85709-413-1)

Breadmaking, 2nd Edition
(ISBN 978-0-85709-060-7)

Cereal Grains: Assessing and Managing Quality
(ISBN 978-1-84569-563-7)

Flour and Breads and Their Fortification in Health and Disease Prevention
(ISBN 978-0-12-380886-8)

The Chorleywood Bread Process
(ISBN 978-1-85573-962-8)

Woodhead Publishing Series in Food Science, Technology, and Nutrition: Number 303

Steamed Breads

Ingredients, Processing and Quality

Sidi Huang

Diane Miskelly

ELSEVIER

AMSTERDAM • BOSTON • CAMBRIDGE • HEIDELBERG
LONDON • NEW YORK • OXFORD • PARIS • SAN DIEGO
SAN FRANCISCO • SINGAPORE • SYDNEY • TOKYO
Woodhead Publishing is an imprint of Elsevier

WP
WOODHEAD PUBLISHING

Woodhead Publishing is an imprint of Elsevier
The Officers' Mess Business Centre, Royston Road, Duxford, CB22 4QH, United Kingdom
50 Hampshire Street, 5th Floor, Cambridge, MA 02139, United States
The Boulevard, Langford Lane, Kidlington, OX5 1GB, United Kingdom

Library of Congress Cataloging-in-Publication Data
A catalog record for this book is available from the Library of Congress

British Library Cataloguing-in-Publication Data
A catalogue record for this book is available from the British Library

ISBN: 978-0-08-100715-0 (print)
ISBN: 978-0-08-100724-2 (online)

Publisher: Nikki Levy
Acquisition Editor: Rob Sykes
Editorial Project Manager: Karen Miller
Production Project Manager: Jason Mitchell
Designer: Victoria Pearson

Typeset by TNQ Books and Journals

Contents

Woodhead Publishing Series in Food Science, Technology, and Nutrition

1 **Chilled foods: A comprehensive guide**
Edited by C. Dennis and M. Stringer
2 **Yoghurt: Science and technology**
A. Y. Tamime and R. K. Robinson
3 **Food processing technology: Principles and practice**
P. J. Fellows
4 **Bender's dictionary of nutrition and food technology, Sixth edition**
D. A. Bender
5 **Determination of veterinary residues in food**
Edited by N. T. Crosby
6 **Food contaminants: Sources and surveillance**
Edited by C. Creaser and R. Purchase
7 **Nitrates and nitrites in food and water**
Edited by M. J. Hill
8 **Pesticide chemistry and bioscience: The food-environment challenge**
Edited by G. T. Brooks and T. Roberts
9 **Pesticides: Developments, impacts and controls**
Edited by G. A. Best and A. D. Ruthven
10 **Dietary fibre: Chemical and biological aspects**
Edited by D. A. T. Southgate, K. W. Waldron, I. T. Johnson and G. R. Fenwick
11 **Vitamins and minerals in health and nutrition**
M. Tolonen
12 **Technology of biscuits, crackers and cookies, Second edition**
D. Manley
13 **Instrumentation and sensors for the food industry**
Edited by E. Kress-Rogers
14 **Food and cancer prevention: Chemical and biological aspects**
Edited by K. W. Waldron, I. T. Johnson and G. R. Fenwick
15 **Food colloids: Proteins, lipids and polysaccharides**
Edited by E. Dickinson and B. Bergenstahl
16 **Food emulsions and foams**
Edited by E. Dickinson
17 **Maillard reactions in chemistry, food and health**
Edited by T. P. Labuza, V. Monnier, J. Baynes and J. O'Brien

18 **The Maillard reaction in foods and medicine**
Edited by J. O'Brien, H. E. Nursten, M. J. Crabbe and J. M. Ames
19 **Encapsulation and controlled release**
Edited by D. R. Karsa and R. A. Stephenson
20 **Flavours and fragrances**
Edited by A. D. Swift
21 **Feta and related cheeses**
Edited by A. Y. Tamime and R. K. Robinson
22 **Biochemistry of milk products**
Edited by A. T. Andrews and J. R. Varley
23 **Physical properties of foods and food processing systems**
M. J. Lewis
24 **Food irradiation: A reference guide**
V. M. Wilkinson and G. Gould
25 **Kent's technology of cereals: An introduction for students of food science and agriculture, Fourth edition**
N. L. Kent and A. D. Evers
26 **Biosensors for food analysis**
Edited by A. O. Scott
27 **Separation processes in the food and biotechnology industries: Principles and applications**
Edited by A. S. Grandison and M. J. Lewis
28 **Handbook of indices of food quality and authenticity**
R. S. Singhal, P. K. Kulkarni and D. V. Rege
29 **Principles and practices for the safe processing of foods**
D. A. Shapton and N. F. Shapton
30 **Biscuit, cookie and cracker manufacturing manuals, Volume 1: Ingredients**
D. Manley
31 **Biscuit, cookie and cracker manufacturing manuals, Volume 2: Biscuit doughs**
D. Manley
32 **Biscuit, cookie and cracker manufacturing manuals, Volume 3: Biscuit dough piece forming**
D. Manley
33 **Biscuit, cookie and cracker manufacturing manuals, Volume 4: Baking and cooling of biscuits**
D. Manley
34 **Biscuit, cookie and cracker manufacturing manuals, Volume 5: Secondary processing in biscuit manufacturing**
D. Manley
35 **Biscuit, cookie and cracker manufacturing manuals, Volume 6: Biscuit packaging and storage**
D. Manley
36 **Practical dehydration, Second edition**
M. Greensmith
37 **Lawrie's meat science, Sixth edition**
R. A. Lawrie

38 **Yoghurt: Science and technology, Second edition**
A. Y. Tamime and R. K. Robinson
39 **New ingredients in food processing: Biochemistry and agriculture**
G. Linden and D. Lorient
40 **Benders' dictionary of nutrition and food technology, Seventh edition**
D. A. Bender and A. E. Bender
41 **Technology of biscuits, crackers and cookies, Third edition**
D. Manley
42 **Food processing technology: Principles and practice, Second edition**
P. J. Fellows
43 **Managing frozen foods**
Edited by C. J. Kennedy
44 **Handbook of hydrocolloids**
Edited by G. O. Phillips and P. A. Williams
45 **Food labelling**
Edited by J. R. Blanchfield
46 **Cereal biotechnology**
Edited by P. C. Morris and J. H. Bryce
47 **Food intolerance and the food industry**
Edited by T. Dean
48 **The stability and shelf-life of food**
Edited by D. Kilcast and P. Subramaniam
49 **Functional foods: Concept to product**
Edited by G. R. Gibson and C. M. Williams
50 **Chilled foods: A comprehensive guide, Second edition**
Edited by M. Stringer and C. Dennis
51 **HACCP in the meat industry**
Edited by M. Brown
52 **Biscuit, cracker and cookie recipes for the food industry**
D. Manley
53 **Cereals processing technology**
Edited by G. Owens
54 **Baking problems solved**
S. P. Cauvain and L. S. Young
55 **Thermal technologies in food processing**
Edited by P. Richardson
56 **Frying: Improving quality**
Edited by J. B. Rossell
57 **Food chemical safety, Volume 1: Contaminants**
Edited by D. Watson
58 **Making the most of HACCP: Learning from others' experience**
Edited by T. Mayes and S. Mortimore
59 **Food process modelling**
Edited by L. M. M. Tijskens, M. L. A. T. M. Hertog and B. M. Nicolaï
60 **EU food law: A practical guide**
Edited by K. Goodburn
61 **Extrusion cooking: Technologies and applications**
Edited by R. Guy

62 **Auditing in the food industry: From safety and quality to environmental and other audits**
Edited by M. Dillon and C. Griffith
63 **Handbook of herbs and spices, Volume 1**
Edited by K. V. Peter
64 **Food product development: Maximising success**
M. Earle, R. Earle and A. Anderson
65 **Instrumentation and sensors for the food industry, Second edition**
Edited by E. Kress-Rogers and C. J. B. Brimelow
66 **Food chemical safety, Volume 2: Additives**
Edited by D. Watson
67 **Fruit and vegetable biotechnology**
Edited by V. Valpuesta
68 **Foodborne pathogens: Hazards, risk analysis and control**
Edited by C. de W. Blackburn and P. J. McClure
69 **Meat refrigeration**
S. J. James and C. James
70 **Lockhart and Wiseman's crop husbandry Eighth edition**
H. J. S. Finch, A. M. Samuel and G. P. F. Lane
71 **Safety and quality issues in fish processing**
Edited by H. A. Bremner
72 **Minimal processing technologies in the food industries**
Edited by T. Ohlsson and N. Bengtsson
73 **Fruit and vegetable processing: Improving quality**
Edited by W. Jongen
74 **The nutrition handbook for food processors**
Edited by C. J. K. Henry and C. Chapman
75 **Colour in food: Improving quality**
Edited by D. MacDougall
76 **Meat processing: Improving quality**
Edited by J. P. Kerry, J. F. Kerry and D. A. Ledward
77 **Microbiological risk assessment in food processing**
Edited by M. Brown and M. Stringer
78 **Performance functional foods**
Edited by D. Watson
79 **Functional dairy products, Volume 1**
Edited by T. Mattila-Sandholm and M. Saarela
80 **Taints and off-flavours in foods**
Edited by B. Baigrie
81 **Yeasts in food**
Edited by T. Boekhout and V. Robert
82 **Phytochemical functional foods**
Edited by I. T. Johnson and G. Williamson
83 **Novel food packaging techniques**
Edited by R. Ahvenainen
84 **Detecting pathogens in food**
Edited by T. A. McMeekin

85 **Natural antimicrobials for the minimal processing of foods**
Edited by S. Roller
86 **Texture in food, Volume 1: Semi-solid foods**
Edited by B. M. McKenna
87 **Dairy processing: Improving quality**
Edited by G. Smit
88 **Hygiene in food processing: Principles and practice**
Edited by H. L. M. Lelieveld, M. A. Mostert, B. White and J. Holah
89 **Rapid and on-line instrumentation for food quality assurance**
Edited by I. Tothill
90 **Sausage manufacture: Principles and practice**
E. Essien
91 **Environmentally-friendly food processing**
Edited by B. Mattsson and U. Sonesson
92 **Bread making: Improving quality**
Edited by S. P. Cauvain
93 **Food preservation techniques**
Edited by P. Zeuthen and L. Bøgh-Sørensen
94 **Food authenticity and traceability**
Edited by M. Lees
95 **Analytical methods for food additives**
R. Wood, L. Foster, A. Damant and P. Key
96 **Handbook of herbs and spices, Volume 2**
Edited by K. V. Peter
97 **Texture in food, Volume 2: Solid foods**
Edited by D. Kilcast
98 **Proteins in food processing**
Edited by R. Yada
99 **Detecting foreign bodies in food**
Edited by M. Edwards
100 **Understanding and measuring the shelf-life of food**
Edited by R. Steele
101 **Poultry meat processing and quality**
Edited by G. Mead
102 **Functional foods, ageing and degenerative disease**
Edited by C. Remacle and B. Reusens
103 **Mycotoxins in food: Detection and control**
Edited by N. Magan and M. Olsen
104 **Improving the thermal processing of foods**
Edited by P. Richardson
105 **Pesticide, veterinary and other residues in food**
Edited by D. Watson
106 **Starch in food: Structure, functions and applications**
Edited by A.-C. Eliasson
107 **Functional foods, cardiovascular disease and diabetes**
Edited by A. Arnoldi

108 **Brewing: Science and practice**
D. E. Briggs, P. A. Brookes, R. Stevens and C. A. Boulton
109 **Using cereal science and technology for the benefit of consumers: Proceedings of the 12th International ICC Cereal and Bread Congress, 24 – 26th May, 2004, Harrogate, UK**
Edited by S. P. Cauvain, L. S. Young and S. Salmon
110 **Improving the safety of fresh meat**
Edited by J. Sofos
111 **Understanding pathogen behaviour: Virulence, stress response and resistance**
Edited by M. Griffiths
112 **The microwave processing of foods**
Edited by H. Schubert and M. Regier
113 **Food safety control in the poultry industry**
Edited by G. Mead
114 **Improving the safety of fresh fruit and vegetables**
Edited by W. Jongen
115 **Food, diet and obesity**
Edited by D. Mela
116 **Handbook of hygiene control in the food industry**
Edited by H. L. M. Lelieveld, M. A. Mostert and J. Holah
117 **Detecting allergens in food**
Edited by S. Koppelman and S. Hefle
118 **Improving the fat content of foods**
Edited by C. Williams and J. Buttriss
119 **Improving traceability in food processing and distribution**
Edited by I. Smith and A. Furness
120 **Flavour in food**
Edited by A. Voilley and P. Etievant
121 **The Chorleywood bread process**
S. P. Cauvain and L. S. Young
122 **Food spoilage microorganisms**
Edited by C. de W. Blackburn
123 **Emerging foodborne pathogens**
Edited by Y. Motarjemi and M. Adams
124 **Benders' dictionary of nutrition and food technology, Eighth edition**
D. A. Bender
125 **Optimising sweet taste in foods**
Edited by W. J. Spillane
126 **Brewing: New technologies**
Edited by C. Bamforth
127 **Handbook of herbs and spices, Volume 3**
Edited by K. V. Peter
128 **Lawrie's meat science, Seventh edition**
R. A. Lawrie in collaboration with D. A. Ledward
129 **Modifying lipids for use in food**
Edited by F. Gunstone

153 **Case studies in food product development**
Edited by M. Earle and R. Earle
154 **Delivery and controlled release of bioactives in foods and nutraceuticals**
Edited by N. Garti
155 **Fruit and vegetable flavour: Recent advances and future prospects**
Edited by B. Brückner and S. G. Wyllie
156 **Food fortification and supplementation: Technological, safety and regulatory aspects**
Edited by P. Berry Ottaway
157 **Improving the health-promoting properties of fruit and vegetable products**
Edited by F. A. Tomás-Barberán and M. I. Gil
158 **Improving seafood products for the consumer**
Edited by T. Børresen
159 **In-pack processed foods: Improving quality**
Edited by P. Richardson
160 **Handbook of water and energy management in food processing**
Edited by J. Klemeš, R. Smith and J.-K. Kim
161 **Environmentally compatible food packaging**
Edited by E. Chiellini
162 **Improving farmed fish quality and safety**
Edited by Ø. Lie
163 **Carbohydrate-active enzymes**
Edited by K.-H. Park
164 **Chilled foods: A comprehensive guide, Third edition**
Edited by M. Brown
165 **Food for the ageing population**
Edited by M. M. Raats, C. P. G. M. de Groot and W. A. Van Staveren
166 **Improving the sensory and nutritional quality of fresh meat**
Edited by J. P. Kerry and D. A. Ledward
167 **Shellfish safety and quality**
Edited by S. E. Shumway and G. E. Rodrick
168 **Functional and speciality beverage technology**
Edited by P. Paquin
169 **Functional foods: Principles and technology**
M. Guo
170 **Endocrine-disrupting chemicals in food**
Edited by I. Shaw
171 **Meals in science and practice: Interdisciplinary research and business applications**
Edited by H. L. Meiselman
172 **Food constituents and oral health: Current status and future prospects**
Edited by M. Wilson
173 **Handbook of hydrocolloids, Second edition**
Edited by G. O. Phillips and P. A. Williams
174 **Food processing technology: Principles and practice, Third edition**
P. J. Fellows

175 **Science and technology of enrobed and filled chocolate, confectionery and bakery products**
Edited by G. Talbot
176 **Foodborne pathogens: Hazards, risk analysis and control, Second edition**
Edited by C. de W. Blackburn and P. J. McClure
177 **Designing functional foods: Measuring and controlling food structure breakdown and absorption**
Edited by D. J. McClements and E. A. Decker
178 **New technologies in aquaculture: Improving production efficiency, quality and environmental management**
Edited by G. Burnell and G. Allan
179 **More baking problems solved**
S. P. Cauvain and L. S. Young
180 **Soft drink and fruit juice problems solved**
P. Ashurst and R. Hargitt
181 **Biofilms in the food and beverage industries**
Edited by P. M. Fratamico, B. A. Annous and N. W. Gunther
182 **Dairy-derived ingredients: Food and neutraceutical uses**
Edited by M. Corredig
183 **Handbook of waste management and co-product recovery in food processing, Volume 2**
Edited by K. W. Waldron
184 **Innovations in food labelling**
Edited by J. Albert
185 **Delivering performance in food supply chains**
Edited by C. Mena and G. Stevens
186 **Chemical deterioration and physical instability of food and beverages**
Edited by L. H. Skibsted, J. Risbo and M. L. Andersen
187 **Managing wine quality, Volume 1: Viticulture and wine quality**
Edited by A. G. Reynolds
188 **Improving the safety and quality of milk, Volume 1: Milk production and processing**
Edited by M. Griffiths
189 **Improving the safety and quality of milk, Volume 2: Improving quality in milk products**
Edited by M. Griffiths
190 **Cereal grains: Assessing and managing quality**
Edited by C. Wrigley and I. Batey
191 **Sensory analysis for food and beverage quality control: A practical guide**
Edited by D. Kilcast
192 **Managing wine quality, Volume 2: Oenology and wine quality**
Edited by A. G. Reynolds
193 **Winemaking problems solved**
Edited by C. E. Butzke
194 **Environmental assessment and management in the food industry**
Edited by U. Sonesson, J. Berlin and F. Ziegler
195 **Consumer-driven innovation in food and personal care products**
Edited by S. R. Jaeger and H. MacFie

196 **Tracing pathogens in the food chain**
Edited by S. Brul, P. M. Fratamico and T. A. McMeekin
197 **Case studies in novel food processing technologies: Innovations in processing, packaging, and predictive modelling**
Edited by C. J. Doona, K. Kustin and F. E. Feeherry
198 **Freeze-drying of pharmaceutical and food products**
T.-C. Hua, B.-L. Liu and H. Zhang
199 **Oxidation in foods and beverages and antioxidant applications, Volume 1: Understanding mechanisms of oxidation and antioxidant activity**
Edited by E. A. Decker, R. J. Elias and D. J. McClements
200 **Oxidation in foods and beverages and antioxidant applications, Volume 2: Management in different industry sectors**
Edited by E. A. Decker, R. J. Elias and D. J. McClements
201 **Protective cultures, antimicrobial metabolites and bacteriophages for food and beverage biopreservation**
Edited by C. Lacroix
202 **Separation, extraction and concentration processes in the food, beverage and nutraceutical industries**
Edited by S. S. H. Rizvi
203 **Determining mycotoxins and mycotoxigenic fungi in food and feed**
Edited by S. De Saeger
204 **Developing children's food products**
Edited by D. Kilcast and F. Angus
205 **Functional foods: Concept to product, Second edition**
Edited by M. Saarela
206 **Postharvest biology and technology of tropical and subtropical fruits, Volume 1: Fundamental issues**
Edited by E. M. Yahia
207 **Postharvest biology and technology of tropical and subtropical fruits, Volume 2: Açai to citrus**
Edited by E. M. Yahia
208 **Postharvest biology and technology of tropical and subtropical fruits, Volume 3: Cocona to mango**
Edited by E. M. Yahia
209 **Postharvest biology and technology of tropical and subtropical fruits, Volume 4: Mangosteen to white sapote**
Edited by E. M. Yahia
210 **Food and beverage stability and shelf life**
Edited by D. Kilcast and P. Subramaniam
211 **Processed Meats: Improving safety, nutrition and quality**
Edited by J. P. Kerry and J. F. Kerry
212 **Food chain integrity: A holistic approach to food traceability, safety, quality and authenticity**
Edited by J. Hoorfar, K. Jordan, F. Butler and R. Prugger
213 **Improving the safety and quality of eggs and egg products, Volume 1**
Edited by Y. Nys, M. Bain and F. Van Immerseel
214 **Improving the safety and quality of eggs and egg products, Volume 2**
Edited by F. Van Immerseel, Y. Nys and M. Bain

215 **Animal feed contamination: Effects on livestock and food safety**
Edited by J. Fink-Gremmels
216 **Hygienic design of food factories**
Edited by J. Holah and H. L. M. Lelieveld
217 **Manley's technology of biscuits, crackers and cookies, Fourth edition**
Edited by D. Manley
218 **Nanotechnology in the food, beverage and nutraceutical industries**
Edited by Q. Huang
219 **Rice quality: A guide to rice properties and analysis**
K. R. Bhattacharya
220 **Advances in meat, poultry and seafood packaging**
Edited by J. P. Kerry
221 **Reducing saturated fats in foods**
Edited by G. Talbot
222 **Handbook of food proteins**
Edited by G. O. Phillips and P. A. Williams
223 **Lifetime nutritional influences on cognition, behaviour and psychiatric illness**
Edited by D. Benton
224 **Food machinery for the production of cereal foods, snack foods and confectionery**
L.-M. Cheng
225 **Alcoholic beverages: Sensory evaluation and consumer research**
Edited by J. Piggott
226 **Extrusion problems solved: Food, pet food and feed**
M. N. Riaz and G. J. Rokey
227 **Handbook of herbs and spices, Second edition, Volume 1**
Edited by K. V. Peter
228 **Handbook of herbs and spices, Second edition, Volume 2**
Edited by K. V. Peter
229 **Breadmaking: Improving quality, Second edition**
Edited by S. P. Cauvain
230 **Emerging food packaging technologies: Principles and practice**
Edited by K. L. Yam and D. S. Lee
231 **Infectious disease in aquaculture: Prevention and control**
Edited by B. Austin
232 **Diet, immunity and inflammation**
Edited by P. C. Calder and P. Yaqoob
233 **Natural food additives, ingredients and flavourings**
Edited by D. Baines and R. Seal
234 **Microbial decontamination in the food industry: Novel methods and applications**
Edited by A. Demirci and M. O. Ngadi
235 **Chemical contaminants and residues in foods**
Edited by D. Schrenk
236 **Robotics and automation in the food industry: Current and future technologies**
Edited by D. G. Caldwell

237 **Fibre-rich and wholegrain foods: Improving quality**
Edited by J. A. Delcour and K. Poutanen
238 **Computer vision technology in the food and beverage industries**
Edited by D.-W. Sun
239 **Encapsulation technologies and delivery systems for food ingredients and nutraceuticals**
Edited by N. Garti and D. J. McClements
240 **Case studies in food safety and authenticity**
Edited by J. Hoorfar
241 **Heat treatment for insect control: Developments and applications**
D. Hammond
242 **Advances in aquaculture hatchery technology**
Edited by G. Allan and G. Burnell
243 **Open innovation in the food and beverage industry**
Edited by M. Garcia Martinez
244 **Trends in packaging of food, beverages and other fast-moving consumer goods (FMCG)**
Edited by N. Farmer
245 **New analytical approaches for verifying the origin of food**
Edited by P. Brereton
246 **Microbial production of food ingredients, enzymes and nutraceuticals**
Edited by B. McNeil, D. Archer, I. Giavasis and L. Harvey
247 **Persistent organic pollutants and toxic metals in foods**
Edited by M. Rose and A. Fernandes
248 **Cereal grains for the food and beverage industries**
E. Arendt and E. Zannini
249 **Viruses in food and water: Risks, surveillance and control**
Edited by N. Cook
250 **Improving the safety and quality of nuts**
Edited by L. J. Harris
251 **Metabolomics in food and nutrition**
Edited by B. C. Weimer and C. Slupsky
252 **Food enrichment with omega-3 fatty acids**
Edited by C. Jacobsen, N. S. Nielsen, A. F. Horn and A.-D. M. Sørensen
253 **Instrumental assessment of food sensory quality: A practical guide**
Edited by D. Kilcast
254 **Food microstructures: Microscopy, measurement and modelling**
Edited by V. J. Morris and K. Groves
255 **Handbook of food powders: Processes and properties**
Edited by B. R. Bhandari, N. Bansal, M. Zhang and P. Schuck
256 **Functional ingredients from algae for foods and nutraceuticals**
Edited by H. Domínguez
257 **Satiation, satiety and the control of food intake: Theory and practice**
Edited by J. E. Blundell and F. Bellisle
258 **Hygiene in food processing: Principles and practice, Second edition**
Edited by H. L. M. Lelieveld, J. Holah and D. Napper
259 **Advances in microbial food safety, Volume 1**
Edited by J. Sofos

260 **Global safety of fresh produce: A handbook of best practice, innovative commercial solutions and case studies**
Edited by J. Hoorfar
261 **Human milk biochemistry and infant formula manufacturing technology**
Edited by M. Guo
262 **High throughput screening for food safety assessment: Biosensor technologies, hyperspectral imaging and practical applications**
Edited by A. K. Bhunia, M. S. Kim and C. R. Taitt
263 **Foods, nutrients and food ingredients with authorised EU health claims: Volume 1**
Edited by M. J. Sadler
264 **Handbook of food allergen detection and control**
Edited by S. Flanagan
265 **Advances in fermented foods and beverages: Improving quality, technologies and health benefits**
Edited by W. Holzapfel
266 **Metabolomics as a tool in nutrition research**
Edited by J.-L. Sébédio and L. Brennan
267 **Dietary supplements: Safety, efficacy and quality**
Edited by K. Berginc and S. Kreft
268 **Grapevine breeding programs for the wine industry**
Edited by A. G. Reynolds
269 **Handbook of antimicrobials for food safety and quality**
Edited by T. M. Taylor
270 **Managing and preventing obesity: Behavioural factors and dietary interventions**
Edited by T. P. Gill
271 **Electron beam pasteurization and complementary food processing technologies**
Edited by S. D. Pillai and S. Shayanfar
272 **Advances in food and beverage labelling: Information and regulations**
Edited by P. Berryman
273 **Flavour development, analysis and perception in food and beverages**
Edited by J. K. Parker, S. Elmore and L. Methven
274 **Rapid sensory profiling techniques and related methods: Applications in new product development and consumer research**
Edited by J. Delarue, J. B. Lawlor and M. Rogeaux
275 **Advances in microbial food safety: Volume 2**
Edited by J. Sofos
276 **Handbook of antioxidants for food preservation**
Edited by F. Shahidi
277 **Lockhart and Wiseman's crop husbandry including grassland: Ninth edition**
H. J. S. Finch, A. M. Samuel and G. P. F. Lane
278 **Global legislation for food contact materials**
Edited by J. S. Baughan
279 **Colour additives for food and beverages**
Edited by M. Scotter

280 **A complete course in canning and related processes 14th Edition: Volume 1**
Revised by S. Featherstone
281 **A complete course in canning and related processes 14th Edition: Volume 2**
Revised by S. Featherstone
282 **A complete course in canning and related processes 14th Edition: Volume 3**
Revised by S. Featherstone
283 **Modifying food texture: Volume 1: Novel ingredients and processing techniques**
Edited by J. Chen and A. Rosenthal
284 **Modifying food texture: Volume 2: Sensory analysis, consumer requirements and preferences**
Edited by J. Chen and A. Rosenthal
285 **Modeling food processing operations**
Edited by S. Bakalis, K. Knoerzer and P. J. Fryer
286 **Foods, nutrients and food ingredients with authorised EU health claims, Volume 2**
Edited by M. J. Sadler
287 **Feed and feeding practices in aquaculture**
Edited by D. Allen Davis
288 **Foodborne parasites in the food supply web: Occurrence and control**
Edited by A. Gajadhar
289 **Brewing microbiology: design and technology applications for spoilage management, sensory quality and waste valorisation**
Edited by A. E. Hill
290 **Specialty oils and fats in food and nutrition: Properties, processing and applications**
Edited by G. Talbot
291 **Improving and tailoring enzymes for food quality and functionality**
Edited by R. Yada
292 **Emerging technologies for promoting food security: Overcoming the world food crisis**
Edited by C. Madramootoo
293 **Innovation and future trends in food manufacturing and supply chain technologies**
Edited by C. E. Leadley
294 **Functional dietary lipids: Food formulation, consumer issues and innovation for health**
Edited by T. Sanders
295 **Handbook on natural pigments in food and beverages: Industrial applications for improving color**
Edited by R. Carle and R. M. Schweiggert
296 **Integrating the packaging and product experience in food and beverages: A road-map to consumer satisfaction**
Edited by P. Burgess
297 **The stability and shelf life of food, Second edition**
Edited by Persis Subramaniam and Peter Wareing
298 **Multisensory flavor perception: From fundamental neuroscience through to the marketplace**
Edited by Betina Piqueras-Fiszman and Charles Spence

299 **Flavor: From food to behaviors, wellbeing and health**
Edited by Andrée Voilley, Christian Salles, Elisabeth Guichard and Patrick Etiévant
300 **Developing food products for consumers with specific dietary needs**
Edited by Wayne Morley and Steve Osborn
301 **Advances in food traceability techniques and technologies: Improving quality throughout the food chain**
Edited by Montserrat Espiñeira and J. Francisco Santaclara
302 **Innovative food processing technologies: Extraction, separation, component modification and process intensification**
Edited by Kai Knoerzer, Pablo Juliano, and Geoffrey Smithers
303 **Steamed breads: Ingredients, processing and quality**
Sidi Huang and Diane Miskelly

Preface

Chinese steamed bread, or simply steamed bread, as it is termed in this book, has long been a traditional staple food of the wheat-growing areas of northern China. Unlike Western breads, which are baked at high temperatures, steamed breads are cooked in a steamer, giving a product with a thin, soft skin and creamy white color. Steamed bread can be made with and without fillings and we have adopted the terminology of steamed buns to describe the filled product. Steamed breads without fillings but consisting of rolled sheets of different colored and/or flavored doughs are described as steamed rolls. Steamed breads, rolls, and buns have long been part of Asian cuisine, not only in China but also in Southeast Asia, where they have been modified to suit local preferences. The popularity of steamed breads is growing worldwide due mainly to the availability and quality of local and imported products.

The first technical and general interest publications on steamed bread appeared around three decades ago and the volume of research and published literature has been increasing steadily. We must acknowledge all the work undertaken by these dedicated workers, both Asian and Western, in their combined efforts to elucidate the science and technology underlying production of these products. This publication—*Steamed Breads: Ingredients, Processing, and Quality*—is the first book to review, describe, and explain the origin, ingredients, manufacture, quality characteristics, technical innovations, and nutritional value of steamed products. It is timely given the rapid mechanization of production and globalization of the food industry which offers opportunities for manufacturers and unprecedented choice and convenience for customers.

Chapter 1 gives a historical account of steamed bread and buns and compares them with Western bread. The classification, similarities, and differences of the various products are introduced in Chapter 2. Major and minor raw materials and their specifications are detailed in Chapters 3 and 4, along with flour quality requirements for the major product types. Fillings and their preparation are described in Chapter 5. Chapters 6 and 7 address traditional and modern dough processing methods and production of products. Frozen doughs and frozen finished products are a relatively new advance for steamed bread manufacture and are covered in Chapter 8. Laboratory preparation and evaluation of steamed bread are discussed in Chapter 9, along with guidelines for researchers in the field. Product faults resulting in unsalable product can be a source of frustration for manufacturers, and solutions to common problems are offered in Chapter 10. Chapter 11 addresses important issues of product spoilage, food safety, and shelf life. Nutritional issues and health implications are covered in Chapter 12, reflecting the importance of the overall changes to many Asian diets including the prevalence of type 2 diabetes and other diet-related issues. Finally, Chapter 13 takes a look at the prospects for the future for steamed bread.

We hope that this publication will be of benefit to researchers, manufacturers, marketers, and those with a general interest and curiosity in Asian traditional foods. This is a time of rapid growth in the food manufacturing industry which is transforming the art of steamed bread making in the home into a commercial science based on technology and innovation.

We acknowledge the support of Grain Growers Limited (formerly BRI Australia) management and staff and in particular the contributions from Ken Quail, Sonya Richards, and Siem Siah.

Our appreciation also goes to Ms. Sisi Wang, Ms. Chengyi Zhao, Professor Hui Sun, and Mr. Zizhong Yan for their generous assistance and support in the preparation of this manuscript. We also acknowledge the help and advice from Professor Tie Fan, Professor Zhengyu Jin, and Yan Zhang.

We are appreciative of the efforts of Colin Wrigley and Ray Moss who proofread and reviewed selected chapters and to Stan Cauvain, Louis Petrin, and Bronwyn Elliot for discussions and advice during preparation of some of the chapters.

We also value the continuing support, technical expertise, and the editorial direction of Tas Westcott during the preparation of this book.

Finally we acknowledge the Editorial and Management input from the team at Elsevier (Rob Sykes, Karen Miller and Jason Mitchell) who made our book possible.

Sidi Huang
Diane Miskelly

Chapter 1

Introduction to Steamed Bread

1.1 INTRODUCTION

Wheat was grown in the Xinjiang Autonomous Region and Gansu and Henan Provinces in northern China more than 5000 years ago, as has been verified by radiocarbon dating (Dodson et al., 2013). Early Chinese characters found on "oracle bones" (tortoise shells or ox bones used as a form of divination), dating from the Shang Dynasty (1751–1122 BCE), provide written evidence of wheat being widely grown throughout Henan (Fan, 1982). Flour milling commenced during the Warring States period, 475–221 BCE (Chen, 1994, 1995) and became widespread. During this time, the text "*Mozigenzhu*" recording the work of the famous ancient philosopher Mozi contains a reference to "bing," which is the common name of cooked wheaten food at that time (Wang, 1984; Zeng, 2002).

Flour-based foods developed further during the Han Dynasty (206–220 AD) and boiled noodles and sesame seed breads gained popularity. It is believed that steamed breads and buns were first produced during the Three Kingdoms period (220–280 AD). As described in the work "*Shi Wu Ji Yuan*" written by a famous governor, Gao Chen, Prime Minister Zhuguo Lian, in a ceremony before war, ordered the use of meat-filled steamed buns to symbolize human heads (Li, 1998). An early monograph on agriculture, "*Qi Min Yao Shu*," written during the 6th century, detailed methods of making more than 20 varieties of wheaten foods and indicated that fermented dough was used to make a steamed product, which was the earliest steamed bread (Wang, 1985).

It is believed that there were five stages during which significant improvements were made in dough fermentation know-how and application (Wang, 1985). There was a need to provide a "starter" for dough fermentation as there was no available source of "pure" yeast. Therefore residual fermentation products from other fermented foods were initially introduced into the dough to provide the necessary leavening action, flavors, and aromas. This ultimately led to the replacement of the traditional starters with residual dough from previous fermentations.

The first stage (200 AD) was "rice wine fermentation"; the second stage (600 AD) was "sour rice syrup fermentation" as recorded in the book "*Qi Min Yao Shu*." The third stage (1200 AD) was "sourdough fermentation"; the fourth

Steamed Breads. http://dx.doi.org/10.1016/B978-0-08-100715-0.00001-X

stage (1300 AD) was "liquid sourdough plus alkali fermentation"; and the fifth stage (1500 AD) was "improved sourdough fermentation." Sourdough fermentation, which appeared during the Song Dynasty (960–1276 AD), provided the basis for modern fermentation technology. The technology for making steamed bread is almost the same today as during the Yuan Dynasty (1280–1367 AD), when people knew how to use alkali to neutralize the acids produced during fermentation.

The word "*shixin mantou*" (steamed bread without filling) first appeared during the early Qin Dynasty in the novel "*Ru Lin Wai Shi*" (Liu, 2005).

1.2 STEAMED PRODUCT TYPES

China is the world's largest producer of wheat, which is mainly grown in the north, and so it is in the north that steamed bread or *mantou* is a staple food. It is usually eaten hot and can be consumed at all meals. In southern China, which is a rice-growing area, steamed bread is eaten mostly at breakfast. Steamed bread popularity has spread throughout China, and can be found even in remote areas.

Steamed bread was introduced to Japan during the Song and Yuan Dynasties. Lin Jingyin introduced steamed bread to Japan and this is celebrated annually (Ni, 2004). According to the Japanese pronunciation, steamed bun is called "*manju*" and products consumed have various types of fillings such as bean paste, meat, curry, sweet and sour pork, walnut, vegetables, prawn, beef, and shell fish. *Manju* can also be made with rice flour. In Korea, steamed products are the second most popular wheat product after cold buckwheat noodles known as *naengmyon* (Nagao, 1995).

Yum cha, which literally means "drink tea," has spread from Guangdong to most of south east Asia and to Chinatowns in the West. *Dim sum*, which are served during *yum cha*, include small steamed dishes which are usually served in their bamboo steamers. Steamed buns are a popular *dim sum* item.

In Japan, South Korea, Indonesia, Thailand, Malaysia, the Philippines, Vietnam, Hong Kong, Taiwan, and Singapore, it has been reported that 40–50% of wheat consumption is in the form of noodles and steamed bread (Anon, 2004).

Chinese steamed bread is made from wheat flour which is mixed into a dough and then fermented before being cooked by steaming. The most common types of steamed breads are either round or roughly cylindrical in shape. Northern-style steamed breads weigh about 130–150 g and have a white, smooth, shiny skin and no crust. The flavor depends on local preferences, and texture varies from dense to open (Huang and Miskelly, 1991).

As with European bread, where a variety of rolls, buns, and pan bread can be made from the same dough, steamed products such as steamed bread, steamed buns, and twisted rolls can be made from the same dough, although the formulation is different from that used for European bread (Huang, 1999). Steamed products can be made with and without fillings. Some of the different types of steamed products are shown in Fig. 1.1. In this book, steamed bread means steamed product without filling, and steamed bun means steamed product with filling.

FIGURE 1.1 Steamed products in their various forms: steamed bread, buns, and rolls. *Reproduced with permission from Huang, S., 2014. Steamed bread. In: Zhou, W., Hui, Y.H. (Eds.), Bakery Products Science and Technology, second ed. Wiley Blackwell, Chichester, West Sussex, UK, pp. 539–562.*

Traditionally, steamed bread was made at home by hand. With the increase in urbanization and living and working away from home, today almost all steamed bread, buns, and rolls sold in cities are made in factories. Many families have little time to make the products at home and many of the younger generations have lost the art entirely. Mechanization has given rise to batch and continuous production, where automated mixing, sheeting, and molding are used.

1.2.1 Steamed Bread (*Mantou or Moo*)

Steamed bread is called *mantou* or *moo* in China and is made without filling (Fig. 1.2). Steamed bread is more popular in China than in other Asian countries. Steamed bread usually has a round or cylindrical shape. Round northern-style types commonly weigh approximately 130g, whereas the southern-style type is produced in two sizes of 65 or 130g and the Guangdong-style type is smaller, weighing only 25g.

1.2.2 Steamed Buns (*Bao Zi or Bao*)

Steamed buns contain many different fillings (Fig. 1.3). They can be divided broadly into two types: savory and sweet fillings (see Chapter 5).

1.2.3 Steamed Rolls (*Huajuan or Juan*)

Steamed rolls, such as those shown in Fig. 1.4, have condiments such as sesame oil spread between the layers of the dough. Differences in the cutting, stretching,

FIGURE 1.2 Steamed bread.

FIGURE 1.3 Steamed buns.

FIGURE 1.4 Steamed rolls.

rolling, and folding of the dough and of the steaming processes can give diverse shapes and flavors. See Section 7.2.3.

1.2.4 Fancy Steamed Bread

There are numerous types of fancy steamed breads manufactured in China which are consumed as a snack. Popular examples are *rouding mantou* filled with chopped pork and the sweet bread *kaihua mantou* (Fig. 1.5).

1.2.5 *Char Siew Bao*—Guangdong-Style Barbecue Pork Buns

Char siew bao is a unique Guangdong-style steamed bun filled with barbequed pork (*char siew*). It has an extremely soft and fluffy texture and very sweet taste (Fig. 1.6). It is characterized by a split on the top of the bun (which occurs during steaming) and is popular in most parts of Asia and is on the menu for *dim sum*. It is actually similar to steamed cake.

1.2.6 Steamed Cake

There are many types of steamed cakes consumed in Asia which use varied ingredients and processing conditions.

Steamed cake, *fagao* (Fig. 1.7), usually has a much more open texture than steamed bread. It is usually made using wheat flour or blends of wheat flour and

FIGURE 1.5 *Kaihua mantou.*

FIGURE 1.6 *Char siew bao*—Guangdong-style barbecue pork buns.

FIGURE 1.7 Steamed cake (*fagao*).

other grain flours such as corn flour or millet flour, together with water, yeast, and sugar. In the south of China, sweet rice wine is used as the yeast source, resulting in very good flavor. Processing includes mixing to a soft dough, fermentation, molding, proofing, and steaming. *Fagao* has a soft and open texture and sweet taste and is most commonly consumed as a snack or at breakfast.

Steamed egg cake, *zheng zhi dangao* (Fig. 1.8), is made from a batter using flour, eggs, sugar, water, and an aerating agent. The cooked texture is very fine, similar to Western cake, but the color is creamy yellow instead of golden brown. There are many types of steamed egg cake which use additional ingredients such as pandan (leaves or extract from the aromatic screw pine), dates, walnut, and preserved fruits.

FIGURE 1.8 Steamed egg cake, *zheng zhi dangao*.

1.3 SIMILARITIES AND DIFFERENCES BETWEEN BAKED AND STEAMED BREAD

There are many similarities between Western bread and steamed bread. Western bread and steamed bread are both fermented wheaten products. Both are light, porous, flavorsome, and easily digested. Production steps are similar for both, except that Western bread is baked in an oven above 200°C, whereas steamed bread is cooked in a steamer. The differences are described below.

1.3.1 Ingredients

The majority of steamed breads require only wheat flour, water, and yeast or sourdough. Guangdong-style steamed breads may contain additional ingredients such as fat and sugar. In Western bread making, other ingredients such as fat, sugar, salt, milk powder, emulsifier, and bread improvers are commonly used. Western bread requires flour with higher protein content and good protein quality compared with steamed bread (Rubenthaler et al., 1992), while a wider range of protein content is suitable for steamed bread (Huang and Hao, 1994). Flour color is more critical for steamed bread, as it should have a white, smooth skin and white crumb. Chinese steamed bread doughs have 10–15% less water added than Western bread doughs. This results in a characteristic fine, chewy texture, which is preferred by consumers.

1.3.2 Baking

Western bread is baked in an oven at over 200°C, while steamed bread dough is steamed at about 100°C. Steamed bread has a white, soft, and moist skin and has a specific volume between 2.0 and 3.2 mL/g. Bread dough forms a brown and hard crust during baking and has larger specific volume (over 4 mL/g). Western bread has a stronger, baked flavor than steamed bread due to baking at high temperatures.

1.3.3 Lysine Availability and Acrylamide Formation

Baking in a conventional oven results in browning of the crust and, to a lesser extent, darkening of the crumb. This is due to the Maillard reaction in which reducing sugars react with amino groups including lysine. Although the lysine content of steamed and baked bread are similar, lysine availability is higher in steamed bread. This is due to the effect of heating during baking resulting in browning (Tsen et al., 1977).

Steamed products were found to have a lower acrylamide level than those that have a baked crust because more than 99% of the acrylamide comes from the crust (Surdyk et al., 2004). Acrylamide is a neurotoxin, carcinogenic in animals, and a probable human carcinogen (Lignert et al., 2002). Acrylamide has not been detected in steamed bread (Zhang, 2010).

1.3.4 Shelf Life

The steaming process raises the water activity of steamed products and provides suitable conditions for mold and bacterial growth. By contrast, bread dough loses moisture during baking and forms a hard crust. The addition of emulsifier, fat, and vinegar or other preservatives to Western bread can result in a shelf life of up to 5 days. Steamed bread can only be kept for 1 day in summer and 2 days in winter (Faridi and Rubenthaler, 1983).

1.4 SIMILARITIES BETWEEN STEAMED BUNS AND OTHER WHEAT-BASED PRODUCTS

Other wheat-based products with fillings have developed throughout the world. Chinese dumplings (*jiaozi*) and European ravioli are prepared from thin wrappers similar to noodle or pasta dough sheets which are then used to enclose fillings. Other examples of European dumplings include pelmeni, pierogi, and kreplach. Unlike steamed buns, these doughs are not yeasted and the products are cooked by boiling.

Germknoedel is a yeasted dumpling or steamed bun often filled with cooked plums and is a specialty of Austria and Bavaria. A related product, dampfnudeln, is fried to create a crispy base before steaming. The method of cooking is similar to that used for Chinese dumplings, although the characteristics and taste of the products are completely different.

1.5 CURRENT STATUS

In China and other developing parts of Asia, there has been a rapid commercialization of the food and food distribution sectors. In the cities, manual and semimechanized production has been replaced by more efficient automated and semiautomated production. Rising incomes in Asia, urbanization, and lifestyle changes have led to more choice for consumers and a demand for safe, high-quality food.

Traditional sales of steamed products in the streets by bicycle (Fig. 1.9) or trolley have evolved to organized distribution systems (Fig. 1.10) and multiple points of sale:

- high-quality packaged frozen steamed buns and rolls in open freezers in supermarkets;
- food presentations in supermarkets to demonstrate how the steamed products are made and customers can purchase items of choice;
- hot and fresh steamed buns in 7/11 and other convenience stores and shopping centers.

There is still a tradition of trolleys and hawker stalls in the streets or the "wet market" which sell freshly manufactured steamed products.

Special flours for steamed products have been developed in most flour mills across China and many parts of Asia. Frozen or chilled steamed products have become convenience foods, needing only 1–2 min reheating in a microwave oven, providing a fast and convenient meal. Wheat-breeding programs have evolved in China and wheat-exporting countries to target the quality attributes important for end products including bread, steamed bread, noodles, and Chinese dumplings.

In 2008, a large modern central manufacturing plant was established in Shanghai to produce steamed buns. Doughs are molded, proofed, and then blast frozen before delivery to chain shops overnight. The frozen buns are freshly

FIGURE 1.9 Fresh steamed bread is delivered by bicycle in the early morning.

FIGURE 1.10 Steamed bread, buns, and rolls are transported by trucks and refrigerated vehicles.

steamed and sold in the morning. The venture has been a huge success and the shops are now all over Shanghai and have even spread to Hangzhou city, 202 km southwest of Shanghai.

In Thailand, there is large-scale production of frozen steamed buns, which are distributed to supermarkets and convenience stores and also exported to surrounding countries such as Japan, Hong Kong, and Singapore. Franchised outlets are also common. Many new products have been developed, for example, containing healthy plants and cereals, black sesame filling, and custard cream filling (Keeratipibul and Luangsakul, 2012).

In response to healthy eating trends, wholemeal and wholegrain buns are available. In Singapore, wholemeal products with pumpkin, salted mung bean paste, and orange and mango custard have been designed to attract younger coffee shop customers and expand the market (Anon, 2012).

1.6 PRODUCTION OF STEAMED PRODUCTS OUTSIDE ASIA

As described in Section 1.2, steamed products are becoming popular around the world. In Asian supermarkets in Sydney, for instance, there are numerous steamed products of which more than half are imported from China, Hong Kong, Taiwan, Singapore, Malaysia, etc. All of the steamed buns have sweet or vegetable fillings, but not meat because of stringent Australian quarantine regulations. The remainder is manufactured locally. There are three main production avenues:

1. Product manufacturers and small factories that supply steamed products with meat, vegetable, and sweet fillings for *yum cha,* supermarkets, and the food-service sector.
2. Small workshops with fresh steamed products for sale at the front and production area in the back.
3. Restaurants, which produce fresh steamed products daily for customers to buy.

The majority of factories use semiautomated production, with the remainder relying more on manual labor.

1.7 CONCLUSION

The Asian region is undergoing rapid development and current trends, in terms of commercialization, the rise of convenience stores and large supermarkets with a variety of fresh and frozen foods, will continue. Despite the availability of Western foods, preference for locally manufactured traditional foods is still strong, but with a greater variety available than ever before. Consumers are becoming more sensitive to quality, convenience, value, and food safety, thus placing demands on the entire value chain, from wheat breeding to consumer, to deliver nutritious, high-quality, consistent products.

REFERENCES

Anon, 2004. US Farmers Target Australian Wheat Sales. The Age, Melbourne, Australia. http://www.theage.com.au/news/Business/US-farmers-target-Australian-wheat-sales/2004/12/21/1103391753934.html (accessed 11.06.15.).

Anon, 2012. Spring News. Singapore Ministry of Trade and Industry. Business Excellence Secretariat, Singapore, p. 14. www.spring.gov.sg/Resources/Documents/SPRINGnews/FV_Jun12/SPRINGnews_Jun2012.pdf (accessed 24.06.15.).

Chen, S., 1994. Talking about the origin of Chinese steamed bread from the history of wheat culture, wheaten foods, and processing tools in China. Agric. Archaeol. 01, 219–225 (In Chinese).

Chen, S., 1995. Discuss the origin of Chinese steamed bread and other fermented wheaten foods in China. Agric. Archaeol. 03, 218–220 (In Chinese).

Dodson, J.R., Li, X., Zhou, X., Zhao, K., Sun, N., Atahan, P., 2013. Origin and spread of wheat in China. Quat. Sci. Rev. 72, 108–111. http://dx.doi.org/10.1016/j.quascirev.2013.04.021.

Fan, C.Y., 1982. The initial approach to several issues of agricultural development in ancient China. J. Chin. Agric. Hist. 1, 60–68 (In Chinese).

Faridi, H.A., Rubenthaler, G.L., 1983. Laboratory method for producing Chinese steamed bread and effects of formula, steaming and storage on bread starch gelatinization and freshness. In: Proceedings 6th International Wheat Genetics Symposium, Kyoto, Japan, pp. 863–867.

Huang, S., Hao, Q., 1994. Steamed bread processing. In: Lin, Z.J., Wang, G.R. (Eds.), Food Processing and Wheat Quality Improvement. Chinese Agricultural Publishing House, Beijing, China, pp. 359–388 (In Chinese).

Huang, S., Miskelly, D., 1991. Steamed bread – a popular food in China. Food Aust. 43, 346–351.

Huang, S., 1999. Wheat Products: 2. Breads, cakes, cookies, pastries, and dumplings. In: Catharina, Y.W., Liu, K.S., Huang, Y.W. (Eds.), Asian Foods Science & Technology, first ed. Technomic Publishing Co Inc., Lancaster, PA, USA, pp. 71–108.

Huang, S., 2014. Steamed bread. In: Zhou, W., Hui, Y.H. (Eds.), Bakery Products Science and Technology, second ed. Wiley Blackwell, Chichester, West Sussex, UK, pp. 539–562.

Keeratipibul, S., Luangsakul, N., 2012. Chinese steamed buns. In: Hui, Y.H., Özgül Evranuz, E. (Eds.), Handbook of Plant-Based Fermented Food and Beverage Technology. CRC Press, FL, USA, pp. 543–556.

Li, H., 1998. Diet and Culture History of Han and Tang Dynasty. Beijing Teachers, University Publishing House, Beijing, China (In Chinese).

Lignert, H., Grivas, S., Jagerstad, M., Skog, K., Tornqvist, M., Aman, P., 2002. Acrylamide in food: mechanisms of formation and influencing factors during heating of foods. Scand. J. Nutr. 46, 159–172. http://dx.doi.org/10.3402/fnr.v46i4.1456.

Liu, C.H., 2005. Production Techniques of Steamed Products. Chemical Industry Publisher, Beijing, China (In Chinese).

Nagao, S., 1995. Wheat usage in East Asia. In: Faridi, H., Faubion, J.M. (Eds.), Wheat End Uses Around the World. American Association of Cereal Chemists, St. Paul, MN, USA, pp. 167–189.

Ni, H.Y., 2004. The earliest ancestor of steamed bread in Japan – Mr. Lin Jiang-Yin. J. China Healthy Food 1, 69.

Rubenthaler, G.L., Pomeranz, Y., Huang, M.L., 1992. Steamed bread. IV. Negative steamer-spring of strong flours. Cereal Chem. 69, 334–337.

Surdyk, N., Rosen, J., Andersson, R., Aman, P., 2004. Effects of asparagine, fructose, and baking conditions on acrylamide content in yeast-leavened wheat bread. J. Agric. Food Chem. 52, 2047–2051. http://dx.doi.org/10.1021/jf034999w.

Tsen, C.C., Reddy, P.R.K., Gehrke, C.W., 1977. Effects of conventional baking, microwave baking, and steaming on the nutritive value of regular and fortified breads. J. Food Sci. 42, 402–406. http://dx.doi.org/10.1111/j.1365-2621.1977.tb01509.x.

Wang, H., 1984. Mozhijiaoshi. Zhejiang Art Publishing House, Hangzhou, China (In Chinese).

Wang, R., 1985. Chinese Diet in Ancient Time. Light Industry Publishing House, Beijing, China (In Chinese).

Zeng, Z.C., 2002. The legendary story of steamed bread and etymology dialectics. J. Folk Cust. Res. 1, 191–193 (In Chinese).

Zhang, H.Z., 2010. Establishment of Determination Methods for Acrylamide in Foods and Its Application Study (Ph.D. thesis). University of Qingdao, China (In Chinese).

Chapter 2

Classification of Steamed Breads and Buns

2.1 INTRODUCTION

Various forms of steamed products such as steamed bread, buns, and twisted rolls have developed throughout Chinese history (Table 2.1). These were regionally based on differences in geography, weather, agricultural products, and eating habits. Thus different formulations, processing methods, and quality preferences evolved. In northern China, which is a semiarid region, wheat is the main crop and steamed bread is a staple food. Northern-style steamed bread with firm, elastic, and cohesive eating quality is preferred as it provides greater satiety. The steamed bread should also have good chewing properties and a natural wheat flavor.

In southern China, which has a warm climate, rice rather than wheat is the staple food. Popular breakfast foods in the south include southern-style steamed bread and rice porridge. Southern-style steamed breads which are soft and a bit chewy are preferred over the traditional northern-style breads. In the Guangdong region of southern China, rice is the staple food, but a wider choice of food is available from elsewhere in China and abroad. In the Guangdong region, unique steamed products have been developed according to local eating habits and preferences (Shuai, 1998). Here, Western influences have resulted in the consumption of small sweet steamed buns as a dessert, which is not a normal part of a Chinese meal. They are often dipped in sweetened condensed milk before consumption.

Huang and Quail (1996) defined three styles of steamed bread (Fig. 2.1) in China and in East and Southeast Asia: northern, southern, and Guangdong styles. These are shown in Table 2.2, with examples of the different varieties. Style means different ingredients and different quality characteristics. Steamed bread refers to steamed products without fillings and steamed bun to products with fillings. The differences in dough for steamed buns and rolls consumed across China are much less than for steamed breads.

Jizhi mantou, produced and consumed in northern China, is another type of steamed bread which is increasing in popularity. It has a slightly less firm

Steamed Breads. http://dx.doi.org/10.1016/B978-0-08-100715-0.00002-1

TABLE 2.1 Classification of Chinese Steamed Products

Form	Style	Type	Specific Volume mL/g	Spread Ratio	Structure	Eating Quality
Steamed bread	Northern	Traditional	2.0	1.2	Very dense	Very firm, cohesive
		Nontraditional	2.5	1.5	Dense	Firm and cohesive
	Southern		3.0	1.6	Open	Soft, slightly cohesive
	Guangdong		2.6–3.4	1.7	Open	Very soft, not cohesive
Fancy steamed bread					Varies	Varies
Steamed buns and rolls					Open	Soft and slightly cohesive
Guangdong buns					Open	Very soft, not cohesive
Char siew bao					Open	Very soft, fluffy, and not cohesive
Steamed cake					Open	Very soft, fluffy, and not cohesive

Modified from Huang, S., 2014. Steamed Bread. In: Zhou, W., Hui, Y.H. (Eds.), Bakery Products Science and Technology, second ed. Wiley Blackwell, Chichester, West Sussex, U.K., pp. 539–562.

FIGURE 2.1 Three styles of steamed bread, clockwise from left—Guangdong, northern, and southern.

TABLE 2.2 Typical Varieties of Northern-, Southern-, and Guangdong-Style Steamed Bread

Style	Type	Variety	Region
Traditional northern style	Very firm	*Gaozhuang mantou*	Shangdong
	Very firm	*Guanguan mo*	Shaanxi
	Very firm	*Gangzi mantou*	Kaifeng, Henan
Nontraditional northern style	Firm	*Jizhi mantou*	Northern China
Southern style	Soft	*Xiaomian mantou*	Southern China
Guangdong style	Soft	*Naibai mantou*	Guangdong

Based on Anon, 1989. Techniques for Making Wheaten Foods, eleventh ed. Chinese Commercial Publishing House, Beijing, China (In Chinese); Huang, S., Quail, K., 1996. Flour quality guidelines for southern style Chinese steamed bread. In: Wrigley, C.W. (Ed.), Cereals 96 Proceedings of the 46th Australian Cereal Chemistry Conference. Cereal Chemistry Division, Royal Australian Chemical Institute, Melbourne, Australia, pp. 315–318; and Shuai, K., 1998. Selection of Guangdong Dim Sum, second ed. Guangdong Science and Technology Publisher, Guangzhou, China (In Chinese).

texture than the traditional northern-style product but retains some of the firm and chewy characteristics due to the addition of an extra 2–3% water to the dough. *Jizhi mantou* means steamed bread made by machine. Both traditional and *jizhi* types are referred to as northern style.

Su (2005) suggested classifying Chinese steamed breads into soft, firm, and very firm. In this system the soft types includes the Guangdong style and the southern style as defined by Huang and Miskelly (1991) and Huang and Quail (1997). The very firm type was traditional northern style as shown in Table 2.2. The firm types were defined as being intermediate between the traditional northern style and southern style. Both firm and very firm types are consumed in northern China and are equivalent to the northern-style steamed bread described in Huang's system.

2.2 DIFFERENCES IN INGREDIENTS

The ingredients for traditional northern- and southern-style steamed breads are very simple: flour, water, and yeast or sourdough (Huang and Miskelly, 1991). Guangdong-style steamed bread may contain up to 25% sugar and 10% fat (Huang and Quail, 1997). Lard was the traditional source of fat, but has now been replaced by fats derived from palm or soy beans. Raising agents, particularly ammonium bicarbonate, have also been used to increase the softness. Dried milk powder or fresh milk is often used to enhance the flavor and protein content.

2.3 DIFFERENCES IN PHYSICAL PROPERTIES AND EATING QUALITY

Key steamed bread quality parameters include specific volume, spread ratio, skin smoothness, color, and texture.

2.3.1 Specific Volume

Specific volume is defined as the ratio of volume in milliliters to the weight in grams. The specific volume for northern-style steamed bread is less than 2.5, while the specific volume for southern-style steamed bread is around 3.0; and for Guangdong style 2.6–3.4 (Table 2.1).

2.3.2 Spread Ratio

The spread ratio is defined as the ratio of width/height measured in millimeters at the point of maximum width and maximum height. The ratio gives an indication of the profile of the steamed bread. Spread ratios are only applicable for well-rounded products. Steamed breads with a spread ratio of 1.2–1.5 are considered as having a northern-style shape, while a spread ratio of 1.4–1.6 is regarded as a southern-style shape (Table 2.1).

2.3.3 Texture and Eating Quality

There are obvious differences in texture and eating quality between the three styles of steamed bread. Northern style steamed bread is characterized by its dense texture and firm, elastic, and very cohesive eating quality. Southern and Guangdong styles have a more open texture. Consumers in southern China prefer steamed bread which is soft and a bit chewy. Guangdong-style steamed bread is unique, being very sweet and with a very soft and elastic, but not cohesive, eating quality. Quality preferences in China range from less sweet with a cohesive texture in the north through to a sweet, soft but with less cohesive texture in the south. This helps to explain why consumers in Guangzhou do not prefer *gao zhong mantou* and consumers in the north do not prefer the Guangdong style.

2.4 DIFFERENCES IN FLOUR QUALITY REQUIREMENTS

The different styles of steamed bread have different flour quality requirements. These will be discussed in Chapter 3.

2.5 GEOGRAPHIC PREFERENCES

Traditional northern-style steamed bread is most popular in northern China. Taiwanese steamed bread, a sweeter version of northern-style steamed bread, has a very cohesive and elastic texture that the locals refer to as "Q." Southern-style steamed bread, which used to be popular only in southern China, is now popular throughout China as a result of added sugar and fat and the use of the "one step" dough procedure (see Section 6.5.1). Most of the steamed breads now produced in factories in Beijing are less firm than typical traditional northern style (*gaozhuang mantou, gangzi mantou*, and *guanguan mo* in Table 2.2). In fact, the majority of steamed breads now consumed in northern China have a firmness that is between typical northern style and southern style.

Guangdong-style steamed bread is not only popular in the southern Chinese provinces of Guangdong, Fujian, and Hainan, but also in Hong Kong, Singapore, and other Southeast Asian countries. Steamed products consumed in other East and Southeast Asian countries are mainly steamed buns based on the Guangdong-style dough formula (Table 2.3). In Japan and Korea, the winter season is the most popular time for steamed buns. These include pork-filled *nikuman* popular in Japan and Korean *hoppang*, which is filled with red bean paste.

2.5.1 1997 Market Survey

To survey geographic preferences, a number of samples of imported packaged frozen steamed breads and steamed buns were purchased from retail outlets in Sydney, Australia. The products were steamed according to the manufacturers recommendations before assessment. Portions of the cooked crumb were analyzed for fat and sugar content.

Table 2.4 shows the results of the analysis of the steamed breads. One Taiwanese product is closer to the accepted tradition northern style, with low fat and sugar content. The remaining products fall closer to the Guangdong style with their higher expected fat and sugar content.

Samples of cooked crumb were taken for analysis from between the skin and the filling of the steamed buns. No analysis was undertaken on the fillings.

Fat content of the buns varied slightly between regions, with the samples from Singapore and West Malaysia containing a much higher level of fat. Sugar

TABLE 2.3 Typical Examples of Steamed Buns in Asia

Country	Name	Typical Fillings
Hong Kong	*Baozi*	Cream custard, lotus seed paste, red bean paste, pork, prawn, chicken, vegetables
	Char siew bao	*Char siew* pork
Indonesia	*Bapao*	Meat
Japan	*Nikuman*	Pork, vegetables
Japan	*Manju*	Red bean paste (adzuki), pork, curry
Korea	*Hoppang*	Red bean paste
Malaysia	*Pau*	Vegetarian, chicken, curry, red bean, coconut, chocolate
Philippines	*Siopao*	Pork, chicken, boiled egg, sweet fillings
Singapore	*Pau/bao*	Pork, vegetable, mushroom, custard, yam, red bean paste
Taiwan	*Baozi*	Meat, vegetables, curry, red bean paste
Vietnam	*Bánh bao*	Pork, chicken, onions, egg, mushrooms, vegetables, Chinese sausage, boiled egg

TABLE 2.4 Comparison of Fat and Sugar Contents of Commercial Frozen Steamed Breads Sold in Australia and Manufactured in the Southeast Asian Region (1997 Survey)

Origin	Fat (%)[a]	Sugar (%)[a]
Hong Kong	5.2	12.1
Malaysia	1.4	19.1
West Malaysia	7.0	10.1
West Malaysia	7.5	9.7
Shenzhen	1.8	9.1
Singapore	8.2	4.0
Singapore	8.3	4.0
Taiwan	3.0	7.7
Taiwan	2.1	2.1

[a]*As is basis.*

TABLE 2.5 Comparison of Fat and Sugar Contents of the Crumb of Commercial Frozen Steamed Buns Sold in Australia and Manufactured in the Southeast Asian Region (1997 Survey)

Origin	Filling	Fat (%)[a]	Sugar (%)[a]
Hong Kong	Egg custard	2.6	4.2
Hong Kong	Lotus seed paste	2.3	4.1
Hong Kong	Sesame paste	3.1	3.5
Hong Kong	Sesame paste	1.2	8.3
Malaysia		3.3	13.7
Shenzhen	Lotus seed paste	1.8	9.5
Shenzhen	Sesame paste	1.0	10.0
South Korea		2.3	5.0
South Korea	Vegetables	2.9	4.6
South Korea	Red bean paste	3.3	4.4
Thailand	Taro	3.7	3.6

[a]*As is basis.*

contents also varied between regions with the majority of samples with a low fat content having the highest sugar contents (Table 2.5). These results are consistent with the observation that in southern China and Southeast Asia there is a preference for sweeter products. The sample from Malaysia was atypical of the set in that it had a low fat level but very high sugar content.

Overall the fat contents of the steamed bun samples surveyed (Table 2.5) were relatively low compared with the unfilled products surveyed in Table 2.4. This may indicate that manufacturers believe that fat additions ranging from about 2% to 3% are ideal for the manufacture of steamed buns. Sugar addition to the doughs varied with the highest levels found in samples from Shenzhen, Hong Kong, and Malaysia demonstrating the preference for sweeter product from these regions. Overall the data confirm that steamed bread and buns consumed in Southeast Asia are Guangdong style.

2.5.2 2015 Market Survey

A second market survey was undertaken to validate and expand on the findings of the first survey. Frozen packaged steamed breads (*mantou*) were selected from Asian retail outlets in Sydney, Australia, as in the first survey. Steamed buns were not collected as no laboratory product analysis was undertaken. For

TABLE 2.6 Comparison of Fat and Sugar Contents of Commercial Frozen Steamed Breads Sold in Australia and Manufactured in the Southeast Asian Region (2015 Survey)

Origin	Description	Fat (%)[a]	Sugar (%)[a]
China—Henan	Spring onion flower roll	2.5	6.9
China—Henen	Scallion twisted bun	0.6	15.0
China—Henan	Flower roll	0.6	22.5
China—Henan	Scallion twist	1.0	24.0
China	*Mantou*	2.5	5.0
China—Guangzhou	*Mantou*	1.0	12.0
Hong Kong	Steamed roll	9.0	11.0
Malaysia	Mini *mantou*	1.5	8.5
Malaysia	Flower roll	4.0	8.0
Malaysia	*Mantou*	5.3	14.0
Singapore	*Mantou*	1.0	9.6
Taiwan	Shantung bun	2.6	3.5
Taiwan	Two-color *mantou*	3.0	9.0
Taiwan	*Mantou*	4.9	4.9
Taiwan	*Mantou*	0.7	4.0

[a]*As is basis.*

all products, country of origin, product type, nutrition panel information, and ingredient lists were documented. Fat and sugar contents as detailed in the nutrition panel were used to compare with the previous survey. Comparison of Tables 2.4 and 2.6 shows similar trends for fat and sugar addition across regions.

Overall the two surveys show that steamed bread composition and styles have remained relatively static over the time between the two surveys. Fat contents were higher than in traditional northern-style steamed breads but excluding the Hong Kong sample, mostly below 5%. Sugar additions varied markedly within and between regions and between manufacturers. No sample was made using the traditional northern-style formulation with no added sugar. Based on the fat and sugar contents, the samples fall mainly between traditional northern-style steamed bread and Guangdong-style steamed bread reflecting the change in consumer preference to a softer and sweeter product.

2.6 CONCLUSION

Changes in Chinese economic development, urbanization, and life style have resulted in changes in steamed bread manufacture and consumer preference. Traditionally, steamed breads were made at home. However, today the majority of steamed breads, buns, and twisted rolls are made in small- or medium-sized factories in the cities. Some handmade products are still produced in small family workshops and in homes in the countryside. Steamed breads manufactured by the sponge and dough procedure are preferred for flavor and texture.

The popularity of traditional northern-style steamed bread in China is in decline as the younger generations have developed a preference for softer and sweeter products. Factory produced *jizhi mantou* is gaining popularity in northern China as it has a softer texture than traditional northern-style steamed bread. While consumption of southern-style steamed bread is increasing throughout China, in Shanghai, recent investment from Taiwan has seen the emergence of modern factories manufacturing Guangdong-style (with added sugar and fat) frozen steamed breads. All three styles of steamed bread are available in supermarkets in major cities. The current popularity of Guangdong-style steamed breads is predicted to increase due to its impact on the preference of younger generations. In new cities like Shenzhen where the population has migrated from all over China, all three styles of steamed breads are popular and available for purchase.

In Southeast Asian regions there is a preference for softer and sweeter steamed breads and buns with many regional preferences and styles.

REFERENCES

Anon, 1989. Techniques for Making Wheaten Foods, eleventh ed. Chinese Commercial Publishing House, Beijing, China (In Chinese).

Huang, S., Miskelly, D., 1991. Steamed bread – a popular food in China. Food Australia 43, 346–351.

Huang, S., Quail, K., 1996. Flour quality guidelines for southern style Chinese steamed bread. In: Wrigley, C.W. (Ed.), Cereals 96 Proceedings of the 46th Australian Cereal Chemistry Conference. Cereal Chemistry Division, Royal Australian Chemical Institute, Melbourne, Australia, pp. 315–318.

Huang, S., Quail, K., 1997. Effects of ingredients on the quality of Guangdong style steamed bread. In: Tarr, A.R., Ross, A.S., Wrigley, C.W. (Eds.), Cereals 97 Proceedings of the 47th Australian Cereal Chemistry Conference. Cereal Chemistry Division, Royal Australian Chemical Institute, Melbourne, Aust., pp. 81–84.

Huang, S., 2014. Steamed Bread. In: Zhou, W., Hui, Y.H. (Eds.), Bakery Products Science and Technology, second ed. Wiley Blackwell, Chichester, West Sussex, U.K., pp. 539–562.

Shuai, K., 1998. Selection of Guangdong Dim Sum, second ed. Guangdong Science and Technology Publisher, Guangzhou, China (In Chinese).

Su, D.M., 2005. Studies on Classification and Quality Evaluation of Staple Chinese Steamed Bread (Ph.D. thesis). Chinese Agricultural University, Beijing, China (In Chinese, English Abstract).

Chapter 3

Major Ingredients for Dough

3.1 INTRODUCTION

The formula for steamed bread dough varies according to the style. Northern- and southern-style breads are made from a lean formula of wheat flour, water, and yeast or sourdough. Breads made with sourdough may require neutralization with alkali after fermentation. Guangdong style is characterized by the addition of sugar and/or fat. In this chapter, flour, water, sourdough, and yeast are referred to as the major ingredients and their roles in steamed bread are discussed.

Other styles may contain optional ingredients which may include other grain flours (maize, sorghum, millet, buckwheat, and black rice) as well as sugar, fat, salt, milk, milk powder, and emulsifiers. These are discussed in Chapter 4.

3.2 FLOUR

Steamed bread is one of the important end products made from wheat flour in Asia. Depending on the country, both domestic wheats and wheats of international origin—the United States, Canada, Australia, and other exporting countries—are used. Flour millers typically use grists of blended wheats to achieve the required flour qualities for production of the various end products.

China is the largest consumer of wheat in the region (Table 3.1). It is estimated that 45% of the flour produced in China is used for steamed bread (including flat bread types), 40% for noodles, 5% for bread, and 10% for cakes and biscuits (He et al., 2010). There has been a steady increase in wheat consumption in the other traditionally rice-consuming countries shown in Table 3.1. In Japan, South Korea, Indonesia, Thailand, Malaysia, the Philippines, Vietnam, Singapore, Hong Kong, and Taiwan, it has been reported that 40–50% of wheat is used for noodles and steamed bread (Anon, 2004). The remainder is used to make Western bread and cakes, pastries, and biscuits.

3.3 FLOUR QUALITY FOR STEAMED BREAD

Protein and starch are the major components of flour, and together with the water-soluble fractions are important determinants of steamed bread quality

Steamed Breads. http://dx.doi.org/10.1016/B978-0-08-100715-0.00003-3

TABLE 3.1 Estimated Per Capita Consumption of Wheat 2011–13

	Kilograms
China	65.0
Japan	42.0
Korea	47.8
Malaysia	29.9
Indonesia	18.9

USDA accessed online 2015 https://apps.fas.usda.gov.

(Addo et al., 1991; Faridi and Rubenthaler, 1983; Huang and Quail, 1996; Huang et al., 1996, 2000, 2003; Lin, 1983; Rubenthaler et al., 1992). While the quality requirements for flour for Western style bread have received much attention, not as much work has been carried out on steamed bread. However, good processing tolerance and requirement for a high-quality final product are common to both. According to Finney (1978), for bread making, "a flour of good quality should have high water absorption, a medium to medium-long mixing requirement, satisfactory mixing tolerance, good loaf volume potential (considering protein content), and should yield a loaf with good internal crumb grain and colour." Further, Tipples et al. (1982) describe the "ideal" bread making flour as yielding dough with well-balanced dough properties without excessive mixing requirement and tolerant to a range of processing conditions. A similar approach can be applied to steamed bread and this is important with increasing mechanization and globalization of production.

According to the Professional Standard of the People's Republic of China, flour for making steamed bread is divided into Superfine and Common grades (Table 3.2). The major difference between the two grades is that the Superfine Grade has a higher farinograph stability, indicating a flour with stronger dough properties.

3.3.1 Protein Content and Protein Quality

3.3.1.1 Protein Content

There is a good linear relationship between protein content and Western bread quality, whereas this is not the case for steamed bread (Addo et al., 1991; Rubenthaler et al., 1992). The nature of the two breads is different. Steamed bread has a soft and moist skin which does not support the internal structure as well as the relatively hard baked crust of Western-style bread. The different styles of steamed bread have different protein and flour quality requirements.

TABLE 3.2 Physicochemical Index of Wheat Flour for Steamed Bread SB/T 10139-93, Ministry of Commerce, China

	Superfine Grade	Common Grade
Moisture (%)	≤14.0	≤14.0
Ash (dry basis) (%)	≤0.55	≤0.55
Particle size (no. CB36 screen)	All pass	All pass
Wet gluten (%)	25–30	25–30
Farinograph stability (min)	≥3.0	≤3.0
Falling number (s)	≥250	≥250
Sand (%)	≤0.02	≤0.02
Magnetic metals (g/kg)	≤0.003	≤0.003
Smell	No abnormal odor	No abnormal odor

Steamed bread is more sensitive to variations in processing conditions which can result in faults such as wrinkles, brown spots, dimples, and possibly total product collapse.

Depending on the style of steamed bread and the type of wheat used, steamed bread has a window of optimum protein content, outside of which there is no improvement, or even deterioration, in some product quality characteristics. This is discussed later in this chapter.

3.3.1.2 Gluten Protein Composition

High-molecular-weight (HMW) and low-molecular-weight (LMW) glutenin subunits are known to affect flour rheological and end use properties. In a study of Canadian and Chinese wheats, Lukow et al. (1990) found that steamed bread score was positively correlated with a higher HMW *Glu-1* score for flours characterized as containing subunits 5 + 10, which have stronger dough properties. The reverse was true for the weaker wheats with 2 + 12 subunits. Similar results have been reported by Huang et al. (2000), Wang and Li (1998), Zhang (2003) and Zhu et al. (2001). LMW glutenin subunits constitute a smaller proportion of storage protein, but for Chinese steamed bread have a large effect on quality. *Glu-A3*, *Glu-B3*, and *Glu-D3* each contributes to at least one of the external and internal quality characteristics (Xia et al., 2013).

Flour proteins can be separated using dilute solutions of sodium dodecyl sulfate, giving extractable and unextractable fractions. The proportion of unextractable polymeric protein (UPP) is correlated with dough strength (Gupta et al., 1993), and was found to be correlated with the total quality score of Guangdong-style steamed bread (Huang et al., 2000). For northern-style steamed bread, high dough strength, UPP, and %UPP were largely found to be genetic factors, and %UPP was correlated with the total score (Zhang et al., 2008).

Wheat breeders in China are increasingly targeting glutenin composition and %UPP to select varieties with improved protein quality. Higher total polymeric protein and moderately high UPP and %UPP have been identified as significant contributors (Zhang, 2003; Zhang et al., 2008).

3.3.2 Color and Extraction Rate

Consumers prefer steamed bread with a white, shiny smooth skin and white crumb (Huang and Miskelly, 1991). Steamed bread color mainly depends on flour color (Lin et al., 1990), but it is also affected by the addition of alkali which is used to neutralize the fermented dough. Processing parameters can also affect color. These include sheeting (Huang and Hao, 1994) and steaming rate (Chen, 2007). Flour color depends not only on the wheat variety and environment, but also on milling conditions and flour extraction rate. Bran contamination is often measured by measuring flour ash content. High flour ash is associated with darker skin and crumb color and lower total score (Huang et al., 1996). A flour extraction rate of 60–70% with a maximum ash of 0.55% is normally used for steamed bread manufacture in China. A maximum ash content of 0.40–0.45% is required for high-quality steamed buns such as frozen products and those served at banquets. Consumer preference is for a white, rather than yellowish color. Although flour can be bleached with benzoyl peroxide to destroy naturally occurring flour xanthophylls, the resultant very white steamed bread is not liked by consumers.

Debranning prior to milling has positive effects on the volume, volume to weight ratio, height, and structure of steamed bread, but this depends on the wheat type (Sun et al., 2007). A debranning rate of 5% appeared to improve the flour characteristics and steamed bread quality for five Chinese wheat varieties (Lin et al., 2012).

3.3.3 Particle Size of Flour

Flour particle size has an effect on the quality of steamed bread (Zhou et al., 1994). For optimum quality products, flour should be relatively fine (less than 160 μm). Flour with coarser particle size gives steamed bread with poorer color and a less shiny skin, but with marginally higher specific volume and total score. Flour with a very fine particle size gives poorer color, flattened shape, lower specific volume and total score, and sticky eating quality (Table 3.3).

3.3.4 Water Absorption

The different styles of steamed bread have different optimum flour water absorptions. For northern-style steamed bread it is around 60–63% (Huang et al., 1996) and for southern style it is around 58–60% (Huang and Quail, 1996).

TABLE 3.3 Effect of Flour Particle Size on the Quality of Steamed Bread

Flour Sample	Sieve Size	Flour Particle Size (μm)	Flour Whiteness (%)	Flour Water Absorption (%)	Specific Volume (mL/g)	Spread Ratio Score (/15)	Color Score (/10)	Nonsticky Score (/15)	Total Quality Score (/100)
1	54GG	≤318	73.3	60.5	2.33	14.5	8.5	13.5	90.0
2	CB30	≤181	75.0	61.9	2.25	14.5	9.1	13.1	90.8
3	CB36	≤160	76.9	63.2	2.20	14.5	8.3	13.0	88.6
4	CB42	≤123	76.8	62.9	2.19	10.0	7.5	12.0	85.7
5	CB54	≤105	79.2	67.2	2.11	10.0	6.6	10.8	76.1
6	150 mesh		83.4	74.2	1.92	9.0	6.9	10.6	72.4

Reproduced with permission from Huang, S., 2014. Steamed bread. In: Zhou, W., Hui, Y.H. (Eds.), Bakery Products Science and Technology, second ed., Wiley Blackwell, Chichester, West Sussex, UK, pp. 539–562.

Guangdong-style steamed bread with added fat requires 58–60% and without fat requires 54–57%. For *char siew bao*, in which soft wheat is used, it can be concluded on examination of the data in Limley et al. (2013) that water absorption should be around 50%.

3.3.5 Starch

Although starch is the largest component of wheat flour, it is actually third in importance after protein and the water-soluble fractions in terms of its influence on steamed bread quality (Huang and Quail, 1999; Lin, 1983). Starch is composed of the polymers amylose and amylopectin. Steamed bread made from flour with high amylose content gave poorer appearance, smaller volume, poor eating quality, low cohesiveness, and sticky texture (Fang and Wang, 1985; Zhu, 2007). Low amylose content was associated with larger volume, better eating quality, and better textural properties (Fan, 1985; Wang et al., 1998; Zhang and Li, 1993).

Starch gelatinizes when heated in excess water and can be measured using the Rapid Visco Analyzer. In a study of northern Chinese steamed bread, Huang et al. (1996) found that flour peak viscosity had the best correlation with specific volume ($r=0.55, p<.001$), crumb structure ($r=0.52, p<.001$), total score ($r=0.47$, $p<.001$), and eating quality ($r=0.39, p<.001$). Flour viscosity parameters always showed better correlation with steamed bread quality than did the equivalent starch parameters, which was similar to the trend observed for yellow alkaline noodles (Konik et al., 1994).

Low flour swelling volume of 19.5 mL/g (indicating low amylose) was found to give a higher total score for *char siew bao*, and too high swelling volume of 23.4 mL/g resulted in a soft and split product (Limley et al., 2013).

Other starch parameters such as starch damage, starch granule size distribution, and diastatic activity showed minor or no significant correlation with northern- and southern-style steamed bread qualities (Huang et al., 1996; Huang and Quail, 1996).

Zhou et al. (1994) used a ball mill to mechanically damage starch during milling. At less than 50% starch damage, there was a correlation between starch damage and steamed quality scores. At higher than 50% starch damage, there was a decrease in bread volumes. Best product quality was between 6.5% and 9.6% starch damage (Liu et al., 2014).

Starch granule size distribution has been shown to influence end product quality. Wheat starch has a bimodal distribution, with small B granules ranging in size from 2 to 10 μm and large A granules greater than 20–35 μm. Starch granule size distribution significantly affected processing quality of Chinese steamed bread. Best crumb structure and highest score were obtained with reconstituted flour from a Chinese wheat cultivar containing 35% small granules (Guo et al., 2014). These results are similar to those obtained by Park et al. (2005) for straight dough (long fermentation) Western bread.

Starch still has an important role to play in the end product quality as it undergoes changes during cooking processes. Damaged starch increases water absorption and also provides a source of sugars for yeast in the baking process. Starch serves to dilute the protein, provides a suitable surface for the starch–protein union, is flexible during partial gelatinization in a deficiency of water permitting expansion; and by taking water from gluten during gelatinization, in combination with heat causes the protein to set (Sandstedt, 1961).

Steamed bread does not have a firm crust to support its structure and shape, so it collapses more readily when removed from the steamer to ambient temperature. Rubenthaler et al. (1992) refer to this phenomenon as "negative oven spring."

3.3.5.1 Waxy Wheat

Recent interest in the development and applications of waxy genotype wheat flour (little or no amylose) addition to normal flours has benefits such as reduced staling of bread, enhanced freeze–thaw stability, lower gelatinization temperatures, and improved white salted noodle quality. It has been reported that blending waxy wheat flour with regular flour could retain the quality of re-steamed bread after freezing at −18°C (Tan, 2007). There was no effect on product quality at waxy flour addition rates below 10% (Qin et al., 2007).

3.3.6 Other Components

The water-soluble fraction of flour is important to the quality of steamed bread (Huang and Quail, 1999; Lin, 1983).

3.3.6.1 Fractionation and Reconstitution Studies

Using a fractionation and reconstitution technique, Huang and Quail (1999) investigated the effect of various flour components on the quality of Guangdong-style steamed bread. Two flours were used: Leichhardt, a strong hard wheat flour, and a soft biscuit flour from Albany region. Interchanging the water-soluble fractions from the two flours resulted in a decrease in the total score of steamed bread made from reconstituted Leichhardt starch and gluten with Albany water-soluble fraction. However, there was an increase in the specific volume and total quality score of steamed bread made from reconstituted Albany starch and gluten with the Leichhardt water-soluble fraction as shown in Figs. 3.1 (A) and (B). The results also indicated a possible interaction between starch and water-soluble components.

3.3.6.2 Pentosans

In addition to albumins and globulins, the water-soluble fraction contains pentosans. The total pentosan content of wheat flour is about 2–3% and the major component of these nonstarch polysaccharides is arabinoxylans. Water-soluble pentosans comprise approximately 25% of total pentosans (Neukom et al., 1967), although this may be as high as 50% (D'Appolonia and Morad, 1981). Water-soluble pentosans are most frequently conjugated with proteins

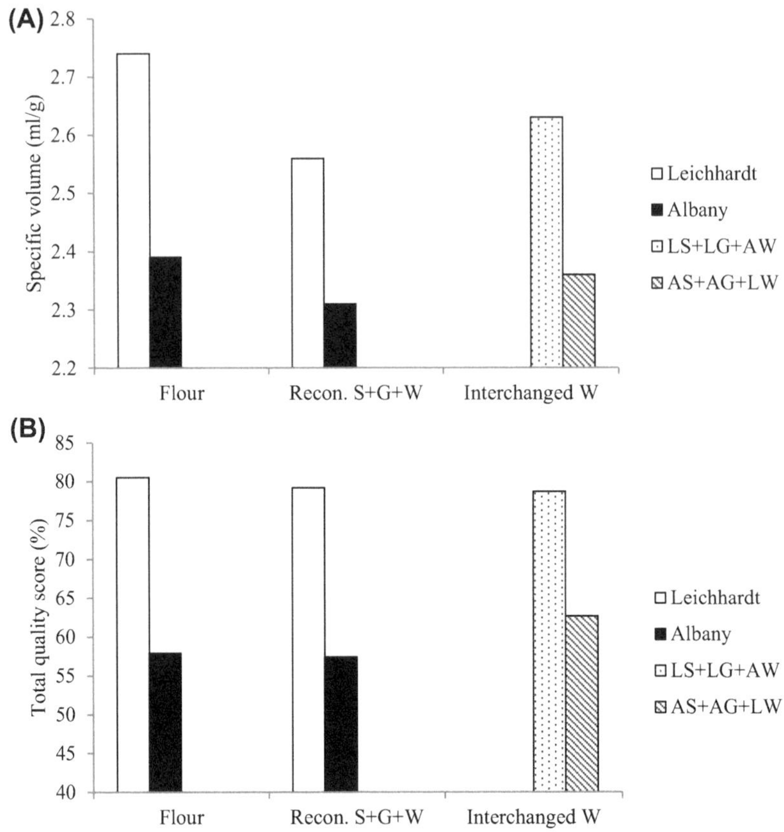

FIGURE 3.1 Differences for specific volume (A) and total quality score (B) of Guangdong-style steamed breads made from flours, reconstituted flours, and reconstituted flours in which water solubles were exchanged. *S*, reconstituted starch; *G*, reconstituted gluten; *W*, water soluble. LS, LG, and LW are the former fractions from Leichhardt; AS, AG, and AW are from Albany. *Modified from Huang, S., 2014. Steamed bread. In: Zhou, W., Hui, Y.H. (Eds.), Bakery Products Science and Technology, second ed., Wiley Blackwell, Chichester, West Sussex, UK, pp. 539–562.*

and possibly other polymers (D'Appolonia and Kim, 1976). They have unique properties, including gel-forming abilities (oxidative gelation) and high water-holding capacity which affects dough mixing properties.

Although there are conflicting reports, water-soluble pentosans seem to improve Western bread-making quality, whereas water-insoluble pentosans are detrimental (D'Appolonia et al., 1970). Pentosans decrease the staling rate, with water-insoluble pentosans having a greater effect than water-soluble pentosans (Kim and D'Appolonia, 1977). However, excessive pentosan levels in flour result in poor baking quality (Kulp, 1968; Neukom, 1976).

It would be expected that some of the above findings for Western bread would be transferable to steamed bread. Dong (2005) studied the effects of added water-soluble pentosans and alkali-soluble pentosans at addition rates of 0.1%, 0.5%, and 1% on the properties of steamed bread. It was found that both water-soluble and alkali-soluble pentosans improved the volume, but did not influence the moisture content of steamed bread crumb. They did increase the softness, but had little effect on elasticity. Addition of either water-soluble or alkali-soluble pentosans inhibited staling, with better results from the water-soluble pentosans.

Water-soluble and insoluble arabinoxylans, in this case extracted from wheat bran, were added to northern-style steamed bread by several authors (Li, 2012; Ma, 2012; Xing, 2012). Xing (2012) found that water-soluble arabinoxylans improved the specific and overall volumes, elasticity, and softness, but had little effect on the color of steamed bread. Furthermore, 2% addition of water-soluble arabinoxylans improved the volume and elasticity, overall sensory quality, and the storage stability, but resulted in uneven development of crumb structure (Li, 2012).

3.3.7 Flour Quality Requirements

For the past 30 years, flour quality requirements for steamed breads have been the focus of research both inside and outside China. This section describes the flour quality requirements for the different styles of steamed bread in China and East and Southeast Asia.

3.3.7.1 Northern Style

Huang et al. (1996) investigated the relationship between flour quality and the quality of northern-style steamed bread using 43 Australian wheats and 6 Chinese wheats. Significant correlations were found between protein content and specific volume ($r=0.49$, $p<.001$) and spread ratio ($r=-0.73$, $p<.001$). When the flour protein was less than 10%, there was a significant positive correlation between protein content and total quality score, as shown in Fig. 3.2. Above 10% protein, the total quality score was more influenced by dough strength parameters—extensograph maximum resistance, development time, stability, mixing tolerance, and extensibility—which were significantly and positively correlated with individual steamed bread quality components. Fig. 3.3 shows the relationship between total score and development time. Overall, dough strength was a more important quality determinant for northern-style steamed bread than protein content. Hard wheat was more suitable than soft wheat, giving higher total scores as shown in Fig. 3.3. It was concluded that best products were obtained using flour with medium protein content and medium to strong dough strength. This was confirmed by Zhang (2003).

An investigation using a set of 11 Australian and 6 Chinese wheats grown under two nitrogen treatments of 0 and 200 kg N/ha was carried out by Zhu et al. (1997). Flour protein content and farinograph dough strength parameters were positively correlated with total score and most of the individual score components. It was confirmed that overstrong flours were not desirable, being

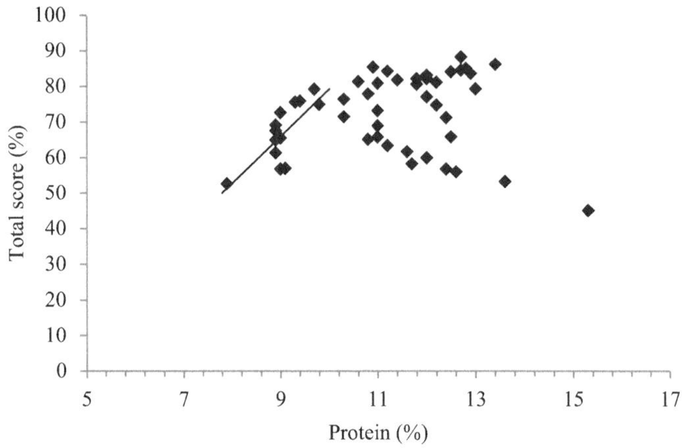

FIGURE 3.2 Relationship between protein content of flour and total quality score of northern-style steamed bread. [Line indicates regression for samples with protein content less than 10% ($R^2=0.67$, $p<.001$)]. *Reproduced with permission from Huang, S., Yun, S.H., Quail, K., Moss, R., 1996. Establishment of flour quality guidelines for northern style Chinese steamed bread. J. Cereal Sci. 24, 179–185. http://dx.doi.org/10.1006/jcrs.1996.0051.*

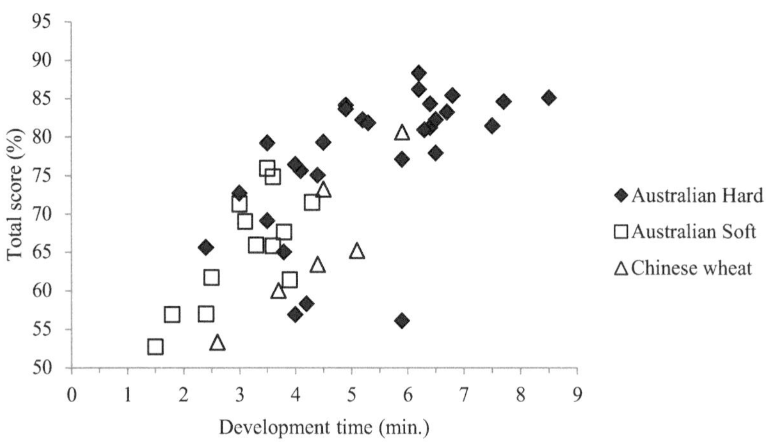

FIGURE 3.3 Relationship between northern-style steamed bread total score and farinograph development time, $r=0.65$, $p<.001$.

associated with steamed bread collapse. In an extension of this work, Zhu et al. (2001) found that for high-protein, high-nitrogen-treated subset of Australian wheats, there was a strong negative correlation between farinograph stability and steamed bread quality, whereas for Chinese wheats, which were generally weaker, the reverse was true. Zhu's conclusion runs counter to that of Huang et al. (1996), but may be due to insufficient samples being used and the impacts of the high nitrogen treatments.

In Su's steamed bread classification system (Table 2.1), the firm and very firm type are both equivalent to northern-style steamed bread. The firm type requires a medium protein content (10.5–12.5%) and for the very firm type medium to high protein content of 10.7–13.5% is suitable (Su, 2005).

Steamed bread in Taiwan has characteristics very close to northern-style steamed bread. The texture is very elastic and has a cohesive eating quality. The flour protein content used in Taiwan is at the higher end of the range and dough properties are strong. To compensate for the higher dough strength, a higher than average steaming rate can be used (Chen, 2007).

It is advisable to have a good balance between flour protein content and dough strength, maintaining an optimum range (Table 3.6) and avoiding high protein content and overly strong dough strength to avoid shrinkage in finished products. For description of this quality fault, see Chapter 10.

3.3.7.2 Southern Style

Early researchers did not differentiate between the various traditional styles of steamed bread. Laboratory formulations and processing procedures reported by Addo et al. (1991), Faridi and Rubenthaler (1983), Lin et al. (1990), Lukow et al. (1990), McMaster and Moss (1989), and Rubenthaler et al. (1992) are considered closer to southern style as they have generally high water additions, and have added sugar and shortening.

There is general agreement that protein content and dough strength play a role in southern-style steamed bread quality (Addo et al., 1991; Huang and Quail, 1996; Lin et al., 1990; Lukow et al., 1990; McMaster and Moss, 1989; Rubenthaler et al., 1990). The optimum flour protein content is about 10%; and above 12% protein, steamed breads exhibit a wrinkled surface and poor color (Lin et al., 1990).

Huang and Quail (1996) found that protein content was correlated with specific volume ($r=0.67$, $p<.001$), spread ratio ($r=-0.60$, $p<.001$), and smoothness ($r=-0.52$, $p<.001$)). Protein content had a stronger effect on the specific volume of southern-style steamed bread compared with northern style. Dough strength also plays a role in determining overall quality as shown in Fig. 3.4, where development time is used as the measure of dough strength. There was also a significant negative correlation between the external smoothness of steamed bread and protein content and between smoothness and flour dough strength. Similar results were reported by Addo et al. (1991) and Rubenthaler et al. (1992).

Western bread and steamed bread have differing requirements for protein content and gluten strength. Addo et al. (1991) made steamed bread and pan bread from 29 hard red winter and hard red spring wheat flours in the protein range 9.3–14.4%; and 34 soft red winter, commercial soft and soft white winter wheats in the protein range 6.7–15.5%. They concluded that protein content was important in the production of acceptable steamed bread quality. High dough strength was important for pan bread, but detrimental for steamed bread. There was no significant correlation between pan bread volume and steamed bread volume.

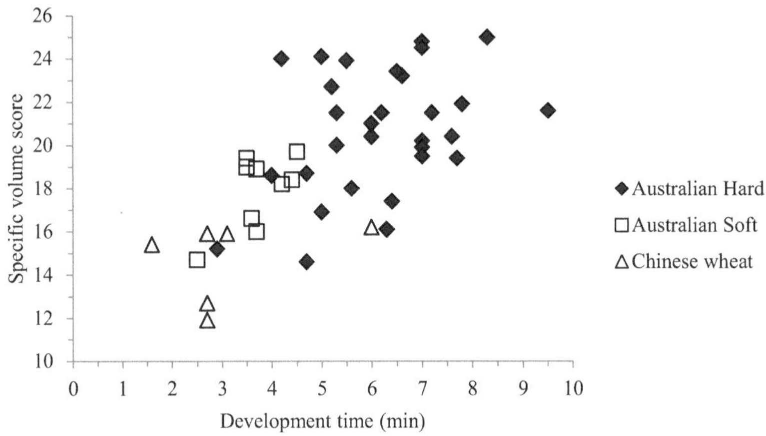

FIGURE 3.4 Relationship between southern-style steamed bread specific volume score and farinograph dough development time, $r = 0.65$, $p < .001$.

Overall, flours with medium protein content and dough strength, a medium to high falling number, and low ash content are recommended for southern-style steamed bread manufacture (Table 3.6). No single test can clearly define the flour quality requirements for southern-style steamed bread. However, hard wheat was found to be superior to soft wheat (Huang and Quail, 1996).

3.3.7.3 Guangdong Style

Guangdong-style steamed bread has evolved from a staple food into a popular snack or dessert after meals and includes ingredients in a range of combinations compared with northern- and southern-style steamed breads. A variety of flours of different quality are used, together with a high level of sugar, but fat is optional. Fat and sugar have a significant effect on the quality of Guangdong-style steamed bread (see Sections 4.4 and 4.6). Different formulations are used in different regions and this may account for the reported variation in flour quality.

Protein content is important for Guangdong-style steamed bread quality. Similar protein effects to those reported for northern and southern styles were found when extracted gluten and starch were added to a base flour (Fig. 3.5). As the protein content increased from 7.8% to 10.8%, the total quality score rapidly increased in a linear fashion up to about 10.5%, after which the rate of increase slowed (Huang and Quail, 1999).

To clarify the flour quality requirements, a consumer test was conducted in Guangzhou, China, using the flours shown in Table 3.4 (Huang et al., 2000). Albany was a pilot milled low protein Australian soft wheat flour; White magnolia, a commercial Guangdong flour for steamed bread manufacture; and D8062, a pilot milled Australian hard wheat flour.

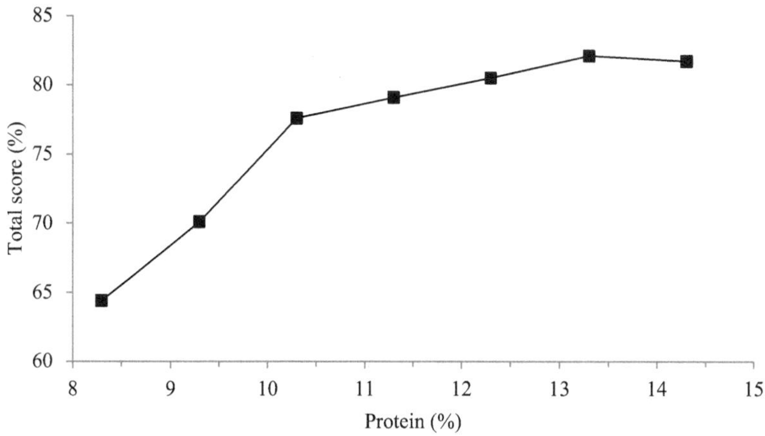

FIGURE 3.5 Relationship between total quality score of Guangdong-style steamed bread and protein content of flour. *Reproduced with permission from Huang, S., 2014. Steamed bread. In: Zhou, W., Hui, Y.H. (Eds.), Bakery Products Science and Technology, second ed., Wiley Blackwell, Chichester, West Sussex, UK, pp. 539–562.*

Steamed bread was made with 20% sugar, and combinations with and without fat as detailed in Table 3.5. There were significant differences in preference between the four steamed breads. The steamed bread made from the hard wheat D8062 with sugar and fat addition had the highest score and was the most preferred by consumers. The steamed bread made from the Albany soft wheat flour with sugar and no added fat had the lowest score and was the least preferred. There was little difference in score or preference between the steamed breads made from White magnolia with or without fat addition. These results indicate that the best Guangdong-style steamed bread is made from flour with medium dough strength and medium protein content, and that weak dough strength is not desirable. Similar requirements for medium dough strength for Guangdong-style steamed bread and buns were reported by Shuai (1998).

3.3.7.4 Summary of Flour Quality Requirements for the Different Styles

Flour quality requirements for three styles of steamed bread are summarized in Table 3.6. It should be pointed out that this is only a general guide when using the optimized processing procedures of Huang et al. (1993, 1998). Variations from the recommendations may exist when different processing procedures are used.

3.3.7.5 Fancy Steamed Bread

Flour with medium protein content and medium dough strength is most suitable for *kaihou mantou* and *rouding mantou* (Anon, 1989).

TABLE 3.4 Flour Data for Guangdong-Style Steamed Bread Preference Testing

			Farinograph		Extensograph	
Flour	Protein (%)	Starch Damage (%)	Water Absorption (%)	Development Time (min)	Maximum Resistance (BU)	Extensibility (cm)
Albany	8.1	4.2	52.3	1.8	150	17.2
White magnolia	8.9	6.4	54.2	1.2	342	14.2
D8062	11.1	8.7	63.9	7.7	490	20.4

Reproduced with permission from Huang, S., 2014. Steamed bread. In: Zhou, W., Hui, Y.H. (Eds.), Bakery Products Science and Technology, second ed., Wiley Blackwell, Chichester, West Sussex, UK, pp. 539–562.
BU, Brabender units.

TABLE 3.5 Results of Consumer Test on the Eating Quality of Guangdong-Style Steamed Bread

Flour	Formula[a]	Score[b]	Preferred Rate (%)
Albany	B	3.6[c]	4.4
White magnolia	A	5.4[d]	22.2
White magnolia	B	5.1[d]	18.9
D8062	A	6.0[e]	54.4

[a]*Formula A: 20% sugar, 4% fat; formula B: 20% sugar, no fat.*
[b]*Average score of 90 measurements.*
[c,d,e] Figures followed by different letters are significantly different at $p < .01$.
Reproduced with permission from Huang, S., 2014. Steamed bread. In: Zhou, W., Hui, Y.H. (Eds.), Bakery Products Science and Technology, second ed., Wiley Blackwell, Chichester, West Sussex, UK, pp. 539–562.

TABLE 3.6 Flour Quality Requirements for Chinese Steamed Bread (14% Moisture Basis)

			Guangdong	
Style	Northern	Southern	With Fat	Without Fat
Protein (%)	10–12	9–11	9–11	8–9
Ash (%)	<0.55	<0.55	<0.55	<0.55
Falling number (s, 5 g)	>230	>230	>230	>230
Starch damage (%)	<9.0	<9.0	<9.0	<9.0
Water absorption (%)	60–63	58–60	58–60	55–57
Development time (min)	4–5	3–4	3–4	2

3.3.7.6 Steamed Buns and Rolls

Traditional southern-style steamed bread dough made from flour with medium protein content and medium dough strength is suitable for the production of steamed buns and steamed rolls. The dough for Guangdong-style steamed bread is suitable for the production of Guangdong buns, using flour with a low protein content and medium dough strength when fat is not included in formula (Shuai, 1998).

3.3.7.7 Char Siew Bao—*Guangdong-Style Barbecue Pork Buns*

The flour quality requirements for *char siew bao* are very different from other styles of steamed breads and buns in Guangdong. Limley et al. (2013) made a systematic study of barbecue pork filled buns (*char siew bao*). It was found that all bun quality parameters except crumb springiness were affected by protein content and that the relationship was quadratic rather than linear. Ideal flours were soft wheat flours with 7.5–8.0% protein content, weak dough strength, moderate to high extensibility, and low flour swelling volume.

3.3.7.8 Steamed Cake

As described previously, *fagao* has an open and soft texture. The flour quality requirements are similar to those for southern-style steamed bread. When a blend of wheat flour and corn flour is used, the ratio of corn flour should be controlled, otherwise the cake volume will be reduced and the texture will be tough.

Flour with low protein content and weak dough strength should be used for steamed egg cakes (Shuai, 1998).

3.3.7.9 Product Optimization by Flour Blending

As indicated earlier, flour with too high a protein content and dough strength could result in shrinkage and other quality faults. The use of a blend of high-protein hard strong wheat with soft weak wheat has been suggested (Lin et al., 1990). Bean et al. (1990) found that the strong, tough, and bucky characteristics of gluten of hard white wheat Klasic could be modified by blending with Anza-type wheat, which has weak, extensible gluten. Steamed bread made from the 50/50 blend gave a satisfactory score and dough was easy to process.

Zhang (2003) reported on the blending of Kansas hard red wheat flour Jagger (protein content 11.8%) with US soft (7.2% protein content) and Chinese hard winter YM 18 (9.7% protein content) wheat flour at ratios from 75:25 to 50:50. Jagger has overstrong dough properties which cause shrinkage after steaming and unsatisfactory steamed bread appearance. In this instance, the quality of steamed bread made from the blends was comparable to that made from the Chinese standard control.

Flour blending may thus be a useful technique to improve the quality of steamed bread when suitable flour is not available.

3.4 PREHARVEST RAIN-DAMAGED (SPROUTED) WHEAT

Wheat that has suffered preharvest rain damage will begin to germinate in the head. There are various degrees of damage, with grain appearing less bright and softer and sprouted grains may be evident. There is a rapid increase in the level of the starch-degrading enzyme α-amylase, and this can be accompanied by increases in other enzymes such as proteases and lipases. The falling number test is used universally to measure the effect of α-amylase and hence the degree of sprouting. Depending on the grade, most wheat standards specify a minimum falling number such as 250 or 300 s, and rain-damaged wheat is segregated from sound wheat at receival. Rain-damaged wheat makes poor-quality pan bread, flat bread, noodles, and steamed bread (Edwards et al., 1989; Orth and Moss, 1987; Zhou et al., 1994). It was reported by the latter that falling numbers below 200 s gave low steamed bread volume and poor appearance (flattened shape and surface blisters), poor texture, and sticky eating quality. This was confirmed by Huang and Quail (1996) who found that wheat flour with some rain damage (falling number 230 s) had an unacceptable surface smoothness, and those with severe rain damage (falling number 62 s) had a very dull and rough surface, with large blisters, brown spots, flat shape, and a very sticky and inelastic eating quality.

3.5 FLOUR AGING

There are no reports in the literature on the effect of flour aging on the quality of steamed bread. In a preliminary experiment carried out by one of us (Huang), it was found that steamed bread made from flours stored for more than 6 months at ambient temperature deteriorated in quality with a smaller volume, firmer texture, and "jelly spots" on the surface. This is probably a result of physical and chemical changes in flour components during storage. It is known that flour increases in dough strength and pan bread volume decreases with time of storage (Haigh et al., 1997). Flour which has been stored for more than 6 months should not be used for steamed bread as the potential for shrinkage or collapse will increase.

3.6 WATER

Gluten development begins when dry ingredients are mixed with water and the dough begins to form. Correct dough consistency is controlled by water addition which can determine the machinability of the dough and hence affect the final quality of steamed bread. An optimal moisture content is essential to the character of steamed bread, contributing to its softness, flexibility, and texture. The final dough temperature can be adjusted by controlling the temperature of the dough water.

3.6.1 Water Quality

Water used for steamed bread manufacture should be of potable quality and free from objectionable odor, flavor, and color. Many municipal water supplies have been chlorinated to protect them against harmful bacteria and algae.

3.6.2 Hardness

Water hardness is due to the presence of calcium ions, and to a lesser extent, magnesium, iron, and manganese. Water of medium hardness (less than about

150 mg/kg as calcium carbonate) is desirable for steamed bread production as some of the mineral salts have a strengthening effect on gluten and stimulate yeast. He and Lu (2014) claimed that calcium ions could be used in the modification of the gluten network structure and so improve the structure and elasticity of steamed bread. Excessively hard water can cause excessive toughening of the gluten structure. Hard water could negatively affect the whiteness of steamed bread and it was suggested that 0.1% water hardness was appropriate (He and Lu, 2014), although this is considerably higher than most municipal water supplies.

3.6.3 pH

Water with a pH of about 6.5 is the most suitable for steamed bread production because it helps the fermented dough achieve the optimal pH (Liu, 2005). Outside a pH of 6.0–6.5, specific volume, whiteness, and sensory score decrease (He and Lu, 2014).

3.7 SOURDOUGH

Most steamed bread dough is fermented using either yeast or sour "starter" dough kept from the previous day (Huang and Miskelly, 1991). Such starter doughs contain multiple strains of yeasts and lactic acid bacteria, but at much lower levels than Western commercially available sourdough starters. Thirty-one lactic acid bacteria (all *Lactobacillus*) and eight yeast strains were isolated from sour starter dough in commercial steamed bun factories in Thailand (Luangsakul et al., 2009). Traditionally, sour starter dough has been used as it is not only cheap and readily available, but creates a unique flavor as well. It is still used in some steamed bread factories and at home. In Thailand, steamed buns made from starter dough are popular and have a more cake-like and sticky texture than those made from yeast (Keeratipibul and Luangsakul, 2012).

However, sourdough can produce an unpleasant smell if the sourdough fermentation is carried out for extended times and/or at high temperatures. The dough of steamed bread becomes very sour (pH about 3.8) because of the by-products of bacterial lactic acid fermentation. It must be neutralized by the addition of alkali (normally Na_2CO_3 solution) to raise the pH to about 6.2 (Mao and Xu, 1999). This is a critical step for final product quality (see Section 6.5.3).

The effect of spontaneous (naturally occurring) and lactic acid bacteria fermentation in traditional northern-style steamed bread doughs was investigated by Wu et al. (2012). The addition of sourdough starter decreased dough water absorption, development time, and stability, but increased specific volume. Sourdough starters conferred a better taste, and unique volatile compounds were present in steamed bread with different starters (Wu et al., 2012). A total of 89 volatile compounds have been identified in steamed breads made from flour/whole wheat pre-ferments, with ethanol and 3-methy-1-butanol being the most abundant (Kim et al., 2009).

3.8 YEAST

Baker's yeast has largely replaced the use of sourdough or "starters" for fermentation in most fully or semiautomated steamed bread factories as the rate of fermentation is higher and more consistent.

3.8.1 Types of Yeast

There are three main types of yeast used in the manufacture of steamed products in Asia: compressed (fresh) yeast, active dry yeast, and instant active dry yeast. It is expected that bulk liquid cream yeast will begin to be used with the trend to larger manufacturers.

Instant active dry yeast is the most widely used by smaller manufacturers because it is convenient to use. Instant active dry yeast contains small amounts of emulsifier, enabling more rapid hydration and dispersion in the dough. Instant active dry yeast is more expensive by comparison with compressed yeast, due to drying and vacuum packaging after manufacture. Compressed yeast needs to be refrigerated at 0–6°C and has a shelf life of about 4 weeks compared with 12 months for sealed dried yeast. However, once the package is opened, the shelf life is 1–2 weeks, due to moisture absorption and oxidation.

3.8.2 Use of Yeast

Instant active dry yeast does not need to be activated before use. Active dry yeast requires activation according to manufacturer's instructions. Compressed yeast is best suspended in water at about 38°C. Yeast is very sensitive to the changes in temperature. The rate of gas production in dough increases as temperature increases up to an optimum of about 40°C. It is advisable to avoid direct contact of yeast with salt, sugar, or alkali.

3.8.3 Rate of Addition

Consistency of yeast activity is important so that once the usage rate is established, it can be maintained. In China, fresh yeast usage rate is about 1% by flour weight. As the yeast activity decreases, the rate of usage will need to increase. Any increase will be determined by assessing product quality but will be influenced by yeast aging and storage conditions. Less dried yeast is required when used instead of fresh yeast. Dried yeast is used at the rate of 50–100% of the fresh yeast and instant active dried yeast at the rate of 33%.

3.8.4 Use of Salt-Stressed Yeast

A novel process for southern-style steamed bread using salt-stressed baker's yeast has been reported by Yeh et al. (2009). Baker's yeast was stressed in a 7% salt solution prior to dough mixing. It was found that fermentation time was reduced due to increased gas production and product expansion. The texture

TABLE 3.7 Influence of Yeast Stress Time on the Properties of Steamed Bread

Stress Time (min)	Softness (g)	Specific Volume (mL/g)	Spread Ratio
Control	541.35 ± 24.20^{a}	2.89 ± 0.35^{b}	1.74 ± 0.04^{a}
20	534.38 ± 40.10^{a}	3.06 ± 0.20^{a}	1.64 ± 0.07^{b}
40	463.08 ± 30.00^{b}	3.15 ± 0.30^{a}	1.45 ± 0.08^{b}

Values are mean ± SD.
[a,b] Figures followed by different letters are significantly different at $p < 0.05$.
Data from Yeh, L.T., Wu, M.L., Charles, A.L., Huang, T.C., 2009. A novel steamed bread making process using salt-stressed baker's yeast. Int. J. Food Sci. Technol. 44, 2637–2643. http://dx.doi.org/10.1111/j.1365-2621.2009.02096.x.

of the bread made from salt-stressed yeast was softer, the specific volume was higher, and the spread ratio was lower than the control. Sensory panel rating was also higher. The results of this study are shown in Table 3.7.

3.9 CONCLUSION

Flour from a variety of wheat types of different origins is used in the production of steamed bread. Hard wheat flour is more suitable than soft wheat flour for making traditional northern- and southern-style steamed bread. Protein content and quality are the most important flour quality factors. Excessive dough strength should not be used to avoid quality faults. For *char siew bao*, soft wheat flours with 7.5–8.0% protein content, weak dough strength, moderate to high extensibility, and low flour swelling volume give best results.

Further work is needed to identify the most suitable commercial wheat classes for different and emerging styles of steamed products.

REFERENCES

Addo, C., Pomeranz, Y., Huang, M.L., Rubenthaler, G.L., Jeffers, H.C., 1991. Steamed bread II. Role of protein content and strength. Cereal Chem. 68, 39–42.

Anon, 1989. Techniques for Making Wheaten Foods, eleventh ed. Chinese Commercial Publishing House, Beijing, China (In Chinese).

Anon, 2004. US Farmers Target Australian Wheat Sales. The Age, Melbourne, Australia. http://www.theage.com.au/news/Business/US-farmers-target-Australian-wheat-sales/2004/12/21/1103391753934.html (accessed 11.06.15.).

Bean, M.M., Huang, D.S., Miller, R.E., 1990. Some wheat and flour properties of Klasic – a hard white wheat. Cereal Chem. 67, 307–309.

Chen, C.C., 2007. Taiwanese Style Steamed Bread Quality as Influenced by the Gluten Strength and Steam Generation Rate (M.Sc. thesis). University of Dayie, Taiwan (In Chinese, English abstract).

D'Appolonia, B.L., Kim, S.K., 1976. Recent developments on wheat flour pentosans. Baker's Dig. 50 (3), 45–49, 53–54.

D'Appolonia, B.L., Morad, M.M., 1981. Bread staling. Cereal Chem. 58, 186–190.

D'Appolonia, B.L., Gilles, K.A., Metcalf, D.G., 1970. Effect of water-soluble pentosans on gluten-starch loaves. Cereal Chem. 47, 194–204.

Dong, B., 2005. Studies of Effect of Pentosans on the Quality of Steamed Bread (M.Sc. thesis). Henan Industrial University, China, p. 68 (In Chinese, English abstract).

Edwards, R.A., Ross, A.S., Mares, D., Ellison, F.W., Tomlinson, J.D., 1989. Enzymes from rain-damaged and laboratory-germinated wheat. I. Effects on Product Quality. J. Cereal Sci. 10, 157–167. http://dx.doi.org/10.1016/S0733-5210(89)80044-X.

Fan, K., 1985. The determination of amylose content and its effect on the quality of steamed bread. Chin. Grain Oil Technol. 2, 27–28 (In Chinese, English abstract).

Fang, K., Wang, C., 1985. The determination of amylose content and its effect on the quality of food. Chin. Grain Oil Foods 3, 27–28 (In Chinese, English abstract).

Faridi, H.A., Rubenthaler, G.L., 1983. Laboratory method for producing Chinese steamed bread and effects of formula, steaming and storage on bread starch gelatinization and freshness. In: Proceedings 6th International Wheat Genetics Symposium, Kyoto, Japan, pp. 863–867.

Finney, K.F., 1978. Contribution of individual chemical constituents to the functional (bread making) properties of wheat. In: Pomeranz, Y. (Ed.), Cereals '78: Better Nutrition for the World's Millions. American Association of Cereal Chemists, St. Paul, MN, USA, pp. 139–158.

Guo, Q., He, Z., Xiachun, X., Qu, Y., Zhang, Y., 2014. Effects of wheat starch granule size distribution on qualities of Chinese steamed bread and raw white noodles. Cereal Chem. 91, 623–630. http://dx.doi.org/10.1094/CCHEM-01-14-0015-R.

Gupta, R.B., Khan, K., MacRitchie, F., 1993. Biochemical basis of flour properties in bread wheats.1. Effects of variation in the quantity and size distribution of polymeric protein. J. Cereal Sci. 18, 23–41. http://dx.doi.org/10.1006/jcrs.1993.1031.

Haigh, K.J., Suter, D.A.I., Higgins, P., O'Riordan, B., Gras, P., Bekes, F., 1997. Effect of wheat storage conditions on bread quality. In: Tarr, A.W., Ross, A.S., Wrigley, C.W. (Eds.), Cereals 97. Proceedings of the 47th Australian Cereal Chemistry Conference. Royal Australian Chemical Institute, Melbourne, Australia, pp. 47–51.

He, Y.Q., Lu, Q., 2014. Impacts of water properties on the quality of steamed bread. Carpathian J. Food Sci. Technol. 6 (1), 25–31. http://chimie-biologie.ubm.ro/carpathian_journal (accessed 01.06.15.).

He, Z.H., Xia, X.C., Zhang, Y., 2010. Breeding noodle wheat in China. In: Hou, G.G. (Ed.), Asian Noodles: Science, Technology, and Processing. John Wiley & Sons, Inc., Hoboken, NJ, USA, pp. 1–23.

Huang, S., Hao, Q., 1994. Steamed bread processing. In: Lin, Z.J., Wang, G.R. (Eds.), Food Processing and Wheat Quality Improvement. Chinese Agricultural Publishing House, Beijing, China, pp. 359–388 (In Chinese).

Huang, S., Miskelly, D., 1991. Steamed bread – a popular food in China. Food Aust. 43, 346–351.

Huang, S., Quail, K., 1996. Flour quality guidelines for southern style Chinese steamed bread. In: Wrigley, C.W. (Ed.), Cereals 96 Proceedings of the 46th Australian Cereal Chemistry Conference. Cereal Chemistry Division, Royal Australian Chemical Institute, Melbourne, Australia, pp. 315–318.

Huang, S., Quail, K., 1999. The effect of flour components on quality of Guangdong style steamed bread. In: Panozzo, J.F., Ratcliffe, M., Wootton, M., Wrigley, C.W. (Eds.), Cereals 99, Proceedings of the 49th Australian Cereal Chemistry Conference. Cereal Chemistry Division, Royal Australian Chemical Institute, Melbourne, Australia, pp. 197–200.

Huang, S., Betker, S., Quail, K., Moss, R., 1993. An optimised processing procedure by response surface methodology (RSM) for northern-style Chinese steamed bread. J. Cereal Sci. 18, 89–102. http://dx.doi.org/10.1006/jcrs.1993.1037.

Huang, S., Yun, S.H., Quail, K., Moss, R., 1996. Establishment of flour quality guidelines for northern style Chinese steamed bread. J. Cereal Sci. 24, 179–185. http://dx.doi.org/10.1006/jcrs.1996.0051.

Huang, S., Quail, K., Moss, R., 1998. The optimisation of a laboratory processing procedure for southern style Chinese steamed bread. Int. J. Food Sci. Tech. 33, 345–357. http://dx.doi.org/10.1046/j.1365-2621.1998.00166.

Huang, S.H., Quail, K.J., Jin, Y., O'Brien, L., 2000. The role of proteins in determining the quality of Guangdong style Chinese steamed bread. In: Wootton, M., Batey, I.L., Wrigley, C.W. (Eds.), Cereals 2000 Proceedings of the 11th ICC Cereal and Bread Congress and of the 50th Australian Cereal Chemistry Conference. Cereal Chemistry Division, Royal Australian Chemical Institute, Melbourne, Australia, pp. 573–577.

Huang, S., Ruang, R., Quail, K., 2003. Eating quality and flour quality requirements of Guangdong style steamed bread. In: Black, C.K., Panozzo, J.F. (Eds.), Cereals 2003, Proceedings of the 53rd Australian Cereal Chemistry Conference. Cereal Chemistry Division, Royal Australian Chemical Institute, Melbourne, Australia, pp. 83–86.

Huang, S., 2014. Steamed bread. In: Zhou, W., Hui, Y.H. (Eds.), Bakery Products Science and Technology, second ed. Wiley Blackwell, Chichester, West Sussex, UK, pp. 539–562.

Keeratipibul, S., Luangsakul, N., 2012. Chinese steamed buns. In: Hui, Y.H., Özgül Evranuz, E. (Eds.), Handbook of Plant-Based Fermented Food and Beverage Technology. CRC Press, FL, USA, pp. 543–556.

Kim, S.K., D'Appolonia, B.L., 1977. Bread staling studies III. Effect of pentosans on dough, bread and bread staling rate. Cereal Chem. 54, 225–229.

Kim, Y.S., Huang, W.N., Zhu, H.Y., Rayas-Duarte, P., 2009. Spontaneous sourdough processing of Chinese Northern style steamed bread and their volatile compounds. Food Chem. 114, 685–692.

Konik, C.M., Mikkelsen, L.M., Moss, R., Gore, P.J., 1994. Relationships between physical starch properties and yellow alkaline noodle quality. Stärke 46, 292–299.

Kulp, K., 1968. Pentosans of wheat endosperm. Cereal Sci. Today 13, 414–417.

Li, X., 2012. Effects of Water-insoluble Arabinoxylans from Wheat Bran on Quality of Chinese Steamed Mantou (M.Sc. thesis). Henan University of Technology, Zhengzhou, China (In Chinese, English abstract).

Limley, H.A., Crosbie, G.B., Lim, K.K., Williams, H.G., Diepeveen, D.A., Solah, V.A., 2013. Wheat quality requirements for char siew bao made from Australian soft wheat. Cereal Chem. 90, 231–239. http://dx.doi.org/10.1094/CCHEM-11-12-0152-R.

Lin, Z., Miskelly, D.M., Moss, H.J., 1990. Suitability of various Australian wheats for Chinese-style steamed bread. J. Sci. Food Agric. 53, 203–213. http://dx.doi.org/10.1002/jsfa.2740530208.

Lin, Q., Liu, L.N., Bi, Y., Li, Z.G., 2012. Effects of different debranning degrees on the quality of wheat flour and Chinese steamed bread. Food Bioprocess Technol. 5, 648–656. http://dx.doi.org/10.1007/s11947-010-0335-3.

Lin, T.C., 1983. Effects of Processing Methods and Wheat Flours on the Quality of Steamed Bread (M.Sc. thesis) University of Kansas State, Kansas, USA. http://hdl.handle.net/2097/12974 (accessed 20.02.15.).

Liu, C., Li, L., Hong, J., Zheng, X., Bian, K., Sun, Y., Zhang, J., 2014. Effect of mechanically damaged starch on wheat flour, noodle and steamed bread making quality. Int. J. Food Sci. Technol. 49, 253–260.

Liu, C.H., 2005. Production Techniques of Steamed Products. Chemical Industry Publisher, Beijing, China (In Chinese).

Luangsakul, L., Keeratipibul, S., Jindamorakot, S., Tanasupawat, S., 2009. Lactic acid bacteria and yeasts isolated from the starter doughs for Chinese steamed buns in Thailand. LWT Food Sci. Technol. 1404–1412. http://dx.doi.org/10.1016/j.lwt.2009.03.007.

Lukow, O.M., Zhang, H., Czarnecki, E., 1990. Milling, rheological and end-use quality of Chinese and Canadian spring wheat cultivars. Cereal Chem. 67, 170–176.

Ma, S.P., 2012. Effect of Wheat Bran Water-Extractable Arabinoxylans on Properties of Dough and Quality of Chinese Mantou (M.Sc. thesis). Henan University of Technology, Zhengzhou, China (In Chinese with English abstract).

Mao, Y.Y., Xu, C.J., 1999. Study on the optimal pH using the best technique of alkaline addition to the fermented steamed bread dough using sough dough starter. Stud. Chin. Cook. 16 (2), 7–10 (In Chinese).

McMaster, G., Moss, H.J., 1989. Flour quality requirements of staple foods of Asia and the Middle East. In: Salovaara, H. (Ed.), Wheat End-Use Properties: Wheat and Flour Characterisation for Specific End-uses. Proceedings from ICC '89. Yliopistopaino, Helsinki, pp. 547–553.

Neukom, H., Geissmann, T., Painter, T.J., 1967. New aspects of the functions and properties of the soluble wheat-flour pentosans. Baker's Dig. 41 (5), 52–55.

Neukom, H., 1976. Chemistry and properties of the non-starch polysaccharides (NSP) of wheat flour. Lebensm. Wiss. Technol. 9, 143–148.

Orth, R.A., Moss, H.J., 1987. The sensitivity of various products to sprouted wheat. In: Mares, D.J. (Ed.), Fourth International Symposium on Pre-harvest Sprouting in Cereals. Westview Press, Boulder, Colorado, USA, pp. 167–175.

Park, S.H., Chung, O.K., Seib, P.A., 2005. Effects of varying weight ratios of large and small wheat starch granules on experimental straight-dough bread. Cereal Chem. 82, 166–172. http://dx.doi.org/10.1094/CC-82-0166.

Qin, P., Chen, G.S.H., Ma, C.X., 2007. Effect of waxy wheat flour blends on the quality of Chinese steamed bread. Agric. Sci. China 6, 1275–1282 (In Chinese, English abstract).

Rubenthaler, G.L., Huang, M.L., Pomeranz, Y., 1990. Steamed bread. I. Chinese steamed bread formulation and interactions. Cereal Chem. 67, 471–475.

Rubenthaler, G.L., Pomeranz, Y., Huang, M.L., 1992. Steamed bread. IV. Negative steamer-spring of strong flours. Cereal Chem. 69, 334–337.

Sandstedt, R.M., 1961. The function of starch in the baking of bread. Baker's Dig. 35 (3), 36–44.

Shuai, K., 1998. Selection of Guangdong Dim Sum, second ed. Guangdong Science and Technology Publisher, Guangzhou, China (In Chinese).

Su, D.M., 2005. Studies on Classification and Quality Evaluation of Staple Chinese Steamed Bread (Ph.D. thesis). Chinese Agricultural University, Beijing, China.

Sun, L.J., Zhou, G.V., Zhi, G.A., Li, Z.G., 2007. Effects of different milling methods on flour quality and performance in steamed breadmaking. J. Cereal Sci. 45, 18–23. http://dx.doi.org/10.1016/j.jcs.2006.02.004.

Tan, P., 2007. Study on Physicochemical Characteristics of Waxy Wheat Flour and Effect of Its Blends on Quality of Bread, Dry White Chinese Noodle and Chinese Steamed Bread (Ph.D. thesis). Anhui Agricultural University, Hefei, China (In Chinese, English abstract).

Tipples, K.H., Preston, K.R., Kilborn, R.H., 1982. Breadmaking properties of wheat flour. Baker's Dig. 56 (6), 16–20.

Wang, X., Li, H., 1998. Wheat quality attributes affect the quality of steamed bread. J. Seeds 3, 40–42 (In Chinese).

Wang, L.K., Zhao, N.X., Cheng, A.H., Lan, J., 1998. A laboratory processing procedure and quality evaluation system for Chinese steamed bread. J. Chin. Cereals Oils Assoc. 13 (2), 29–32 (In Chinese).

Wu, C., Liu, R.S., Huang, W.N., Rayas-Duarte, P., Wang, F., Vao, V., 2012. Effect of sourdough fermentation on the quality of Chinese northern-style steamed bread. J. Cereal Sci. 56, 127–133. http://dx.doi.org/10.1016/j.jcs.2012.03.007.

Xia, X.C., He, Z.H., Jin, L.H., Wang, L.H., Zhang, Y., Li, G.Y., 2013. Understanding the effects of HMW-GS and LMW-GS on processing quality using near-isogenic lines and development of functional markers for LMW-GS in wheat. In: He, Z.H., Wang, D. (Eds.), Proceedings 11th International Gluten Workshop China, August 12–15, 2012. D.F. International Maize and Wheat Improvement Center (CIMMYT), Mexico, pp. 107–112.

Xing, L.Y., 2012. Effect of Water-Extractable Arabinoxylans from Wheat Flour on Chinese Mantou (M.Sc. thesis). Henan University of Technology, Zhengzhou, China (In Chinese, English abstract).

Yeh, L.T., Wu, M.L., Charles, A.L., Huang, T.C., 2009. A novel steamed bread making process using salt-stressed baker's yeast. Int. J. Food Sci. Technol. 44, 2637–2643. http://dx.doi.org/10.1111/j.1365-2621.2009.02096.x.

Zhang, C.Q., Li, Q.Q., 1993. Effects of wheat quality attributes on the quality of steamed bread. J. Chin. Agr. Sci. 36 (2), 39–46 (In Chinese, English abstract).

Zhang, P., He, Z., Zhang, Y., Xia, X., Chen, D., Zhang, Y., 2008. Association between % SDS-unextractable polymeric protein (%UPP) and end-use quality in Chinese bread wheat cultivars. Cereal Chem. 85, 696–700. http://dx.doi.org/10.1094/CCHEM-85-5-0696.

Zhang, H.X., 2003. Effects of Flour Protein, Dough Rheological Properties, and Flour Extraction Rates on Northern Style Chinese Steamed Bread Making (Ph.D. thesis). Department of Grain Science and Industry College of Agriculture, Kansas State University, p. 143.

Zhou, W., Lin, H., Liu, H., Chao, F., Lu, H., 1994. The relationships between Chinese steam bread and flour quality factor. In: Xie, G.F., Ma, Z.D. (Eds.), Proceedings 94, International Symposium & Exhibition on New Approaches in the Production of Food Stuffs and Intermediate Products from Cereal Grain and Oil Seeds. Beijing, China, pp. 701–709.

Zhu, J., Huang, S., O'Brien, L., Wei, X., Mares, D.J., 1997. In: Tarr, A.R., Ross, A.S., Wrigley, C.W. (Eds.), Cereals 97 Proceedings of the 47th Australian Cereal Chemistry Conference. Cereal Chemistry Division, Royal Australian Chemical Institute, Melbourne, Australia, pp. 272–275.

Zhu, J., Huang, S., Khan, K., O'Brien, L., 2001. Relationship of protein quantity, quality and dough properties with Chinese steamed bread quality. J. Cereal Sci. 33, 205–212. http://dx.doi.org/10.1006/jcrs.2000.0358.

Zhu, F., 2007. Study on the Quality of the Physicochemical Properties of Wheat Flour and Their Relationship with the Quality of the Chinese Steamed Bread (M.Sc. thesis). Wuhan Polytechnic University, Wuhan, China (In Chinese, English abstract).

Chapter 4

Optional Ingredients for Dough

4.1 INTRODUCTION

In addition to the basic ingredients of wheat flour, water, and yeast or sourdough starter, other ingredients are often included to improve finished product quality and add variety. In this chapter, optional ingredients are introduced briefly and their applicability to steamed bread production is discussed. Given the diversity of products available in the marketplace, it is inevitable that commercial addition rates may not necessarily reflect those reported in the literature, which are frequently based on laboratory-scale processing.

4.2 OTHER FLOURS

4.2.1 Blending of Wheat Flour and Other Flours

A recent trend has been the production of steamed bread using blends of wheat flour and other flours (Fig. 4.1). Consumers perceive these products as healthy alternatives perhaps offering nutritional advantages, even though they often have smaller volume, darker color, denser texture, and stronger flavor.

These alternative nonwheat flours include millet, sorghum, sweet potato, buckwheat, oats, and black rice, which can be substituted for wheat flour at rates depending on the grain used (Table 4.1).

4.2.2 Barley Flour

Barley is a source of soluble and insoluble dietary fiber and β-glucans all of which possess a number of reported important health benefits. Different types of barley flours are available ranging from milled whole grain flour to flours with some components removed to flours milled from pearled barley. Each type of flour will therefore differ in functionality and composition.

Incorporation of 10%, 20%, and 30% barley flour decreased dough stability, work input, and extensibility. Increasing levels of barley flour decreased the specific volume, brightness, and whiteness index and increased the firmness and chewiness of Taiwanese-style steamed bread. This study also showed that

Steamed Breads. http://dx.doi.org/10.1016/B978-0-08-100715-0.00004-5

FIGURE 4.1 Steamed bread and rolls made from wheat flour and blends of wheat flour and maize or wheat flour and sorghum.

TABLE 4.1 Maximum Blending Rate of Nonwheat Flours

Nonwheat Flour	% Wheat Flour Substitution
Millet	45
Sorghum	20–30
Buckwheat	30–40
Maize	30
Oat	20
Sweet potato	15

Data from Fu, Y.M., 2006. Studies on the Factors Affecting the Quality of Steamed Bread Made From the Mixture of Wheat Flour and Other Grain Flours (M.Sc. thesis). Henan University of Technology, Zhengzhou, China (In Chinese, English abstract) and Liu, C.H., 2005. Production Techniques of Steamed Products. Chemical Industry Publisher, Beijing, China (In Chinese).

flour milled from pearled six-row barley contains more dietary fiber than barley endosperm flour (Lin et al., 2012).

4.3 DIETARY FIBER

Fiber is an important part of the diet and is considered beneficial in protection against constipation, heart disease, and cancer.

FIGURE 4.2 Steamed bread made from wheat flour with added bran.

FIGURE 4.3 Steamed bread made from wheat flour and taro, vegetable juice, and wheat bran.

Wheat bran is one of the major dietary fiber sources used to improve the fiber content of processed foods. A large number of dietary fiber sources are used in Western bread making, including bran, wholemeal flour, wheat fiber, pulse hulls, maize, and oats. It is becoming common to find steamed bread fortified with wheat bran or wholemeal flour in supermarkets in Asia (Figs. 4.2 and 4.3).

4.3.1 Wheat Bran

Wu et al. (2012) substituted wheat flour with 10%, 20%, and 30% bran and established that optimum addition was 20%. Increasing the bran content resulted in a firmer texture. Sensory acceptance and elasticity were improved by increasing water addition to 54% or higher, together with addition of shortening and emulsifier [sodium stearoyl-2-lactylate (SSL)]. Further improvements during steaming were achieved by increasing the output of the steam generator.

The particle size of the bran can affect the quality of steamed bread. Small to medium-sized particles are the most suitable. Acceptable quality steamed bread can be produced with 20% medium bran substitution of flour (Bao, 2010).

4.3.2 Barley Spent Grain

Barley spent grain, a by-product of the brewing industry, is high in cellulose, hemicellulose, and protein. Cao (2010) reported that using soluble and insoluble fiber from barley spent grains, with 3% addition of total fiber (ratio of soluble to insoluble fiber at 1:2) improved the textural and sensory quality of northern-style steamed bread.

4.3.3 Barley Hull and Flaxseed Hull Extract

Barley and flaxseed hulls are rich in phenolics and other bioactive compounds. When alkalized extracts were added to Chinese steamed bread, there was an increase in total phenolics, and antioxidant activity increased by between 34.5% and 90.7% indicating that these ingredients have potential as functional food ingredients (Hao and Beta, 2012). Further evidence from animal studies or clinical trials is needed on barley hull and flaxseed hull extracts for food applications to demonstrate safety and efficacy in health promotion and their effect on bread quality.

4.3.4 Soybean Hull

Xiao et al. (2005) reported that addition of 3% soybean hull to steamed bread dough could improve the quality including specific volume, spread ratio score, texture, and elasticity.

4.3.5 Resistant Starch

Resistant starch is considered a form of dietary fiber and has been used because of its digestive health benefits (Zhu, 2014). Fu et al. (2010) found that addition of resistant starch resulted in deterioration of steamed bread quality. However, they reported that the addition of 10% resistant starch to high dough strength flour or 5% addition of resistant starch to medium dough strength flour produced steamed bread with acceptable quality. It was also found that the addition of less than 5% resistant starch produced from buckwheat reduced the quality of steamed bread. This would suggest that stronger flours should be used when resistant starch is added.

4.3.6 Green Banana Flour

Green bananas are readily available in Southeast Asia and green banana flour is high in dietary fiber and resistant starch. Noor Aziah et al. (2012) substituted wheat flour with 30% green banana flour and found an increase in water-holding capacity, crude fiber, dietary fiber, and resistant starch but decreased steamed bread volume. Addition of 8% dry gluten to the formulation improved steamed bread specific volume to an acceptable level, although the crumb was darker.

4.4 SUGAR

Sugar (sucrose) is added to steamed bread and steamed bun doughs primarily for the sweetening effect now required for consumer acceptance. As well as producing sweetness, the sugar replaces some of the flour in the formula and hence has an effect on the dough consistency and on the texture of the final product. Sugar may be added in either solid or liquid form. Granular sugar can be supplied in a number of grades usually specified by particle size distribution. As the sugar needs to dissolve in the dough, granular sugars of different grades may need different water additions and mixing procedures. Addition of liquid sugar is less problematical as its water content is known and allowances can be made easily in production.

Sugar can be used to supplement extended fermentation in flour with a low diastatic activity. Huang and Quail (1997) showed that sugar levels of 8–10% are optimum for Guangdong-style steamed bread and effectively improved the specific volume and total score (Fig. 4.4). At 10% addition, maximum volume is achieved after the first 4h of fermentation (He et al., 2008; Huang and Quail, 1997). Above 10% addition, quality was adversely affected, although surveys of commercial products showed products could be successfully manufactured with

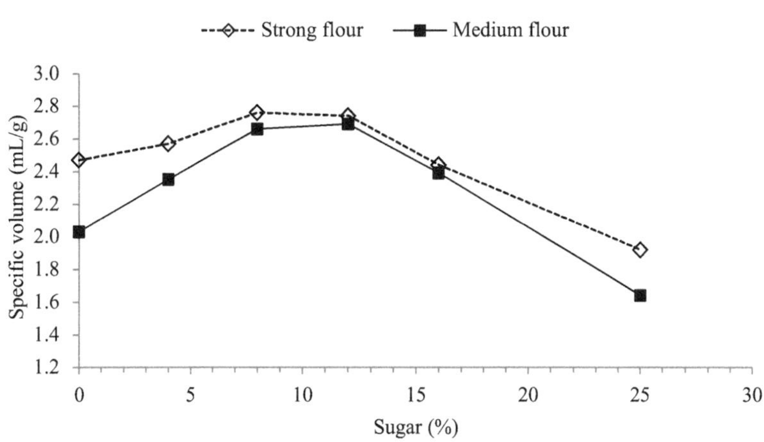

FIGURE 4.4 Effect of the levels of sugar addition on the specific volume of Guangdong-style steamed bread.

addition levels in excess of 20%. At low levels (up to 3%), there were negligible changes in external or internal appearance, volume, width to height ratio, or texture (Kruger et al., 1992).

4.5 SALT

Salt is generally added to foods to enhance flavor, and Western bread without salt can be tasteless. Salt also acts to strengthen the gluten and to reduce the stickiness of dough and thus improve handling properties. However, salt retards fermentation and hence the rate of carbon dioxide production.

Unlike Western bread, salt is not commonly used in steamed bread dough formulations. When it is used, the addition rates are below 0.5% by flour weight. Salt is usually added to the savory filling for steamed buns and/or spread on the extended dough layers for steamed rolls.

Addition of salt can improve the skin color of Guangdong-style steamed bread (Huang and Quail, 1997), but this depends on the flour type. For a medium soft flour (protein content 8.4%, farinograph water absorption 54.1%, and maximum resistance 390 BU), addition of up to 1% salt improved the skin color as shown in Fig. 4.5. For a strong flour (protein content 11.1%, farinograph water absorption 63.9%, and maximum resistance 490 BU), there was less improvement in skin color up to 0.5% addition, and at higher levels over 1% the skin color was badly affected. Similar results were reported by Kruger et al. (1992). With both strong and medium flour, crumb softness was improved at salt addition levels between 0.25% and 0.5%.

The total score of steamed bread made from the strong flour was significantly reduced at salt levels over 0.5% addition, while total score of steamed bread made from the medium flour was significantly increased at 0.25% salt but decreased over 0.5% salt.

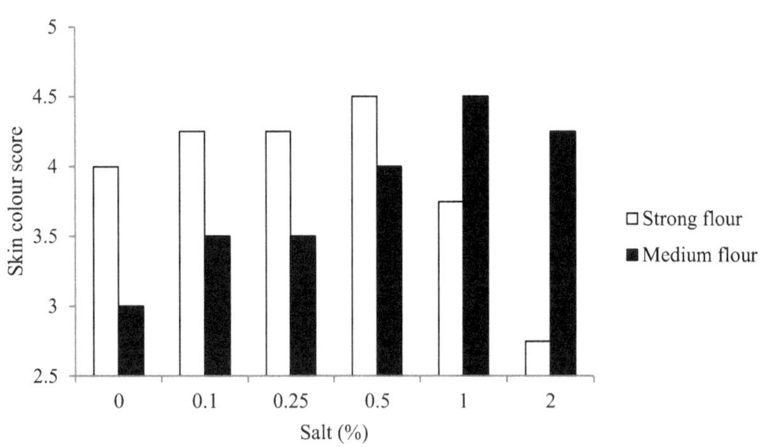

FIGURE 4.5 Effect of salt level on the skin color score of Guangdong-style steamed bread.

Other reported effects of salt addition include a reduction in water absorption. An addition of 2% salt decreases water absorption of steamed bread dough by 3% (Liu, 2005).

4.6 FATS AND EDIBLE OILS

Fats play an important role in steamed bread manufacture as they contribute flavor and texture and are capable of modifying the properties of doughs and finished products. The generally accepted distinction between fats and oils is that fats are solid, whereas oils are liquid, at room temperature. Fats and oils are mixtures of triglycerides each of which has a unique melting point. Because fats do not have sharp melting points, it is possible to have liquid fat and fat crystals coexisting at a given temperature. At very low temperatures, all the fat will be solid and at high temperatures all the fat will be liquid. Determination of the solid to liquid ratio at a particular temperature can be measured using nuclear magnetic resonance spectroscopy. The method directly and accurately measures the solid fat content (SFC) of the sample. Edible oils are not generally considered suitable for use in Western breads and related products as they are liquid at dough processing temperatures. Baking functionality of fats is related to their SFC profile and it is the solid phase that is important in delivering the so-called shortening action. Many commercial shortenings are manufactured from a liquid blend that is then subjected to a special chilling and shearing process to provide the ideal liquid to solid ratio. These products are semisolid at production temperatures and provide optimal shortening effect.

Palm oil from Southeast Asia and soybean oil mainly from China are the most commonly used fats in steamed bread manufacture.

Pomeranz et al. (1991) investigated the role of shortenings on the quality of steamed bread made from soft wheat flour. It was found that addition of 2% partially hydrogenated vegetable shortening, soybean or corn oil improved total volume and softness compared with a control with no added fat. A positive shortening response was obtained even for defatted flour, which is not the case for Western bread. Li and Chen (1999) found that the addition of lard improved volume, color, softness, and shelf life of steamed bread.

For a strong flour, softness of Guangdong-style steamed bread increased significantly with addition of up to 8% vegetable shortening, but at even higher additions, softness decreased (Huang and Quail, 1997). For a medium strength flour, there was a significant increase in softness up to an addition of 2% but higher additions did not further improve softness (Fig. 4.6).

The effect of fat on the total score of Guangdong-style steamed bread is shown in Fig. 4.7. For strong flour, the total score was significantly increased by up to 4% fat addition. Over 8% addition, a significant decrease was observed. For medium flour, the total score was significantly increased only up to 2% fat addition and there was no significant change over 2%. Significant effects on specific volume and elasticity were also observed. Other steamed bread quality

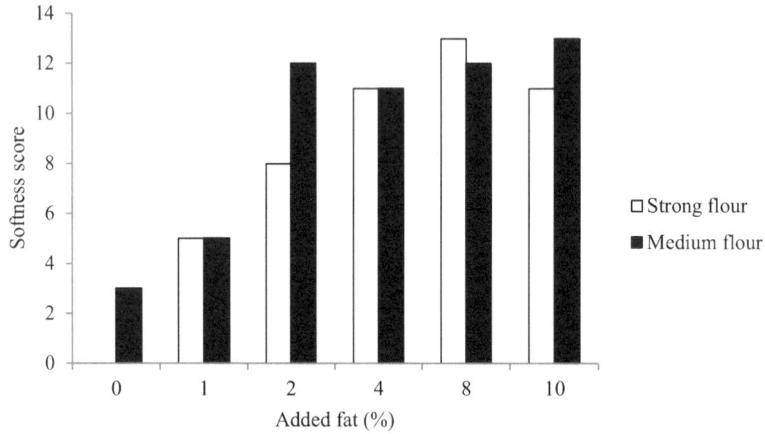

FIGURE 4.6 Effect of fat addition on the softness score of Guangdong-style steamed bread (note softness score for strong flour at zero fat addition was zero).

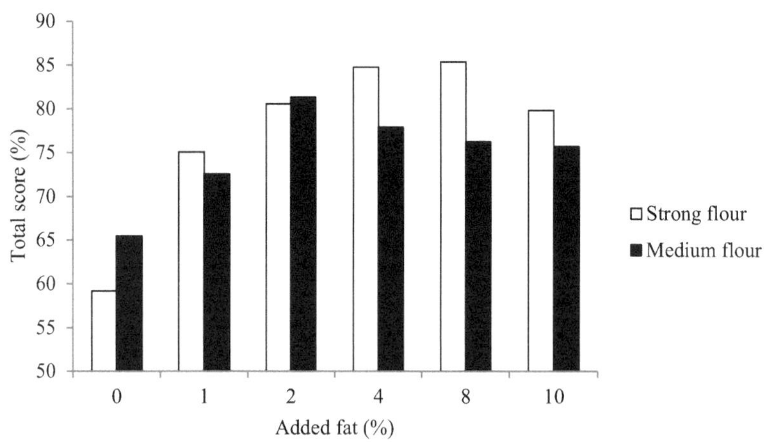

FIGURE 4.7 Effect of fat addition on total score of Guangdong-style steamed bread.

attributes were unaffected. The optimum level of fat addition for Guangdong-style steamed bread quality from this work is 2–4%. These results are consistent with those of Pomeranz et al. (1991).

4.7 ALKALI

Alkaline salts are commonly added to Asian yellow alkaline noodle doughs. Salts such as sodium carbonate, potassium carbonate, and ammonium bicarbonate, or various mixtures known as lye water are used because they raise dough pH, toughen the dough, increase paste viscosity, and develop the characteristic yellow color of yellow alkaline noodles. These alkaline salts are also used to neutralize sourdough fermentations raising the dough pH from about 3.8 to 6.2–6.7. Dough which is underneutralized will collapse or shrink during steaming, as the gluten has become tough and inextensible. If the pH of the dough is above 7.2, the steamed bread will have an undesirable yellow color. There is also evidence from noodle studies that the retention of B group vitamins, particularly riboflavin and thiamin, is lower under alkaline processing conditions (Bui et al., 2013).

4.8 LEAVENING AGENTS

4.8.1 Baking Powder

Baking powders are chemical leavening agents containing sodium bicarbonate, commonly known as baking soda or bicarb soda, plus one or more leavening acids. The chemical leavening acids can be slow or fast acting. Slow-acting baking powders containing, for example, sodium acid pyrophosphate or glucono-δ-lactone, release most of the carbon dioxide gas at steaming or baking temperatures. Fast-acting baking powders containing, for example, calcium triphosphate as the acidulent, release carbon dioxide during mixing at room temperature as well as at steaming or baking temperatures. Commercially available leaveners containing both slow- and fast-acting acidulents are known as double acting and are used in special circumstances where an initial release of carbon dioxide is required before the product is baked.

4.8.2 Ammonium Bicarbonate

Ammonium bicarbonate is a common leavening agent used in the baking industry and has been reported to be used in Guangdong-style steamed breads. It releases carbon dioxide and ammonia gas when the dough temperature reaches 40°C (Lai and Lin, 2006). If used at the correct addition rate, it significantly improves steamed bread color and volume (Yu, 1990). In addition, it can function as a buffering agent (Liu et al., 2006). Overuse of ammonium bicarbonate can result in ammoniacal taints in the final steamed bread. Ammonium bicarbonate is more commonly used in low-moisture products such as English-style biscuits.

4.8.3 Role of Leavening Agents in Steamed Bread

In Guangdong-style steamed buns, slow-acting baking powder is added as a leavening agent in addition to the yeast. It improves steamed bread softness (Fig. 4.8) when added at 0.1% to medium strength flour but there is no improvement in softness above this level. For strong flour, 0.5% addition is optimal.

Additions up to 1% baking powder improved skin color, but the overall quality was not significantly improved. Bitter flavors are detected over 1% addition (Huang and Quail, 1997).

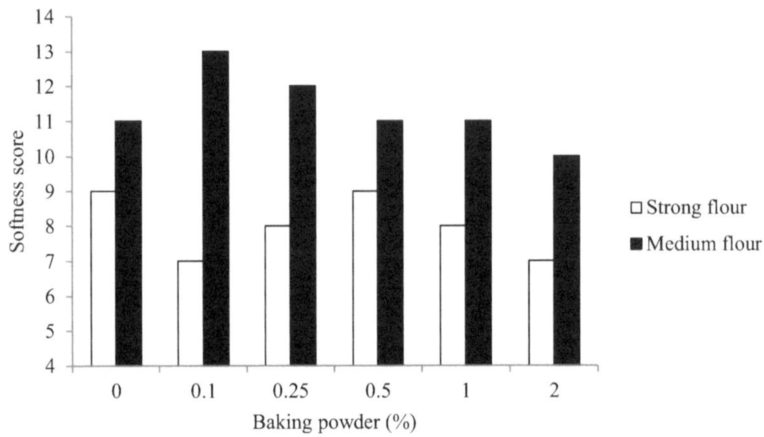

FIGURE 4.8 Effect of baking powder on softness score of steamed bread.

Double-acting baking powder is used for *char siew bao* to give more aeration during the steaming process (Limley, 2008).

Potassium and ammonium aluminum sulfate-based baking powders used to be very popular for use in steamed bread, particularly for Guangdong-style steamed bread and steamed cake (*fagao*) as they impart a very soft and open texture. They were banned from use in food production including steamed bread and buns in China from July 2014 to reduce dietary exposure to aluminum.

4.9 SOY FLOUR

Soy flour has been an ingredient in Western bread making due to its improving effects on dough handling and product quality. It is also a source of high-quality protein, because of its high lysine content compared with other cereal proteins. Enzyme active soy flour is often added at low levels to Western bread where the enzyme lipoxygenase bleaches the xanthophyll pigments, resulting in a whiter crumb. It also functions to produce oxidative dough strengthening.

In steamed bread, the skin and crumb color can be significantly improved by the addition of up to 2% defatted soy flour (Zheng et al., 2006). There was also an improvement in specific volume. However, over 6% addition resulted in poor structure.

Cruickshank (1996) found that the addition of enzyme active soy flour significantly improved the total score of steamed bread, with the highest improvement at 0.6% soy flour addition. At higher levels of addition, specific volume and total score were reduced, and there was a noticeable "beany" soy aroma and flavor. This was supported by Lin (1983) who found steamed bread produced using 7.5% full-fat and defatted soy composite flour was significantly less acceptable than the control. It was also reported that 2% addition of low-temperature defatted soybean flour gave the best quality improvement for steamed bread (Xiao et al., 2005).

4.10 ASCORBIC ACID

Ascorbic acid is used as an ingredient in bread improvers for Western bread. It functions as an oxidant, increasing the strength of the gluten network and producing bread with increased volume, brighter crumb, and better crumb texture and appearance.

Cauvain and Huang (1986) made northern-style steamed bread using a high-intensity dough mixer as used in the Chorleywood Bread Process (CBP). CBP steamed bread with 1% yeast and 75 mg/kg ascorbic acid had a larger volume, more open crumb structure, and firmer crumb than that made using a long fermentation process. Chang et al. (2010) using a planetary-action mixer extended the range of added ascorbic acid from 10 to 90 mg/kg and also concluded that the optimum level of addition was 70 mg/kg.

By contrast, Cruickshank (1996) found that the greatest improvement in total score and the best skin smoothness score of northern-style steamed bread was achieved at ascorbic acid addition of 25 mg/kg. Ascorbic acid additions of 100 and 200 mg/kg produced only a marginal improvement in total score, but skin smoothness was extremely poor. It was concluded that the differences in optimum addition levels were a result of the different flours used. Cruickshank used a farinograph mixer bowl to prepare the dough from a flour with strong dough properties which suited northern-style steamed bread.

4.11 EMULSIFIERS

Emulsifiers are surface active compounds, meaning they are miscible with both lipids and water. They are widely used in many Western bakery products to improve gluten strength, dough handling, elasticity, gas retention, structure, volume, and crumb softness. They can also function to reduce product staling.

Emulsifiers can be categorized according to their charge (ionic/nonionic), solubility, and functionality. Their mechanisms of action can be quite diverse, and emulsifiers interact with both protein and starch components of flour. The correct choice of emulsifier will also depend on the type of fat used. Emulsifier use is controlled by Food Law in most countries.

Among the common emulsifiers are:

- Diacetyl tartaric acid ester of mono- and diglycerides (DATEM): In Western bread, DATEM addition to a dough will strengthen gluten and improve gas retention.
- Calcium stearoyl-2-lactylate and SSL function in stabilizing doughs and improving crumb softness.
- Lecithins, which are naturally occurring polar lipids, are mainly derived from soybeans and play a role in retardation of staling.

Although much work has been carried out in the use of emulsifiers in bakery products, there is limited information on steamed bread. Zhang et al. (2007) compared different DATEM addition rates to Western bread and steamed bread and found that as DATEM addition increased, skin color, elasticity, and crumb structure improved up to an optimum level of 0.10% addition. A similar optimum was found for Western bread. It was found that farinograph development time increased and extensograph extensibility decreased with increasing DATEM addition. However, Ying et al. (2009) found that this response to DATEM depends on flour strength. Weak and strong flours showed improved steamed bread quality, whereas medium-strength flours gave variable results. The effect of DATEM on the quality of steamed breads needs further investigation. Examples of emulsifiers used to improve storage properties of steamed bread are discussed further in Section 11.6.

4.12 ENZYMES

Enzymes are used as flour additives and dough conditioners and are often an ingredient in bread improver formulations for Western bread. Depending on their nature they may reduce mixing time and work input and provide increased dough strength and higher loaf volume. The most common ones used in Western bread are amylases, xylanases, proteases, and lipases. Compared with Western bread, relatively few applications of enzymes in Chinese steamed bread have been reported as yet and their function and potential have not been fully evaluated.

4.12.1 Amylases

Amylases hydrolyze damaged starch and α-amylase may be derived from cereal and fungal and bacterial sources. Fungal α-amylase was found to improve skin and crumb color of steamed bread up to a maximum addition of 50 mg/kg, whereas for maximum specific volume the maximum addition was 30 mg/kg, but at this concentration, the spread ratio score was at its lowest (Bao and Wang, 2008). Nonetheless, the authors concluded that the optimum addition of α-amylase for overall quality improvement was 30 mg/kg.

4.12.2 Xylanases

Xylanases are of particular interest as together with other hemicellulases, they hydrolyze hemicelluloses and reduce dough viscosity, thus improving rheological properties and releasing water to other dough constituents. During the Western bread-making process, they delay crumb formation, enabling larger bread volume (Polizeli et al., 2005).

Su (2005) found that with the addition of both liquid and solid endoxylanase to flour, there was a decrease in farinograph water absorption, dough development time, and stability; and a decrease in mixograph time to peak height. Xylanase reduced the fermentation time by about 30% for southern-style steamed bread and increased specific volume, spread ratio, and softness, but elasticity was unchanged. Xylanase addition also decreased Minolta Brightness (L*) and slightly increased yellowness (b*), therefore the shininess of steamed bread skin was slightly reduced. An increase in specific volume of 20–24% with xylanase addition was also reported by Jiang et al. (2010).

The effect of a heat-tolerant xylanase from a thermophilic bacterium *Geobacillus* sp. *PZH1* was investigated by Wang et al. (2011). The heat-tolerant xylanase significantly elevated steamed bread volume and improved texture and shelf life.

Lin et al. (2010) found that cellulase can improve steamed bread dough rheological properties and steamed bread quality. As cellulase addition increased, so did farinograph water absorption, dough development time, stability, and tolerance index. In addition, extensograph maximum resistance and area increased and extensibility decreased. The best improvement in specific volume, elasticity, and total score was achieved at 0.5% cellulase addition. A much lower optimum level of addition of 60mg/kg was proposed by Lu et al. (2015) using a fuzzy mathematical model. These differing findings may reflect differences in activities of the enzymes used.

The role of enzymes on the storage properties of steamed bread is further discussed in the shelf-life section.

4.13 COMBINATION APPLICATION OF EMULSIFIERS, ENZYMES, AND OTHER ADDITIVES

Blends of different food additives or microingredients are used in commercial Western bread improvers to give a greater effect. Bread improvers are composed of one or more oxidants, blends of enzymes, and can contain emulsifiers, yeast-activating compounds, and soy and malt flour. This enables functional modifications at various stages of the bread-making process and improvements in finished products. Design and composition of bread improvers is a complex task, and most of the expertise resides in the commercial domain.

Some information is available on their application in steamed bread. Single and combination application of emulsifiers, enzymes, and others were investigated by Hu et al. (2003), Ma (2007), Ren and Cheng (2007), Wang and Wang (2006), and Zhang et al. (2014). Details are listed in Table 4.2, together with the main effects on quality. This is an area for further research.

4.14 GUMS AND HYDROCOLLOIDS

Gums and hydrocolloids function as water binders, foam stabilizers, and gel formers in baked products in baking applications such as cakes, batters, muffins, pancakes, and fillings. Some enhance freeze/thaw stability. There is limited information on their applicability to steamed bread, with only one study showing that the addition of sodium alginates (0.2%) or of konjac glucomannan (0.8%) gave Guangdong-style steamed bread a relatively low spread ratio and specific volume, but softer crumb and more resistance to staling on storage (Sim et al., 2011).

TABLE 4.2 Combination Application of Emulsifiers, Enzymes, and Other Additives on Quality of Steamed Bread

Combination	Optimum Concentration Reported (mg/kg)	Effect on Steamed Bread Quality	References
DATEM plus CSL/SSL plus Lipase	3000+ 1000+ 8	Improved color and structure	Wang and Wang (2006)
CSL-SSL plus Lipase	500+ 25	Improved color, appearance, and eating quality. Lower specific volume	Hu et al. (2003)
Endoxylanases plus Ascorbic acid	10+ 8	Improved specific volume and spread ratio score	Zhang et al. (2014)
Amylase plus Xylanase plus Lipase plus Ascorbic acid plus Glucose oxidase	3+ 10+ 4+ 18+ 5	Finer texture. Better elasticity	Ren and Cheng (2007)
Fungal α-amylase plus Lipase plus Xylanase plus Glucose oxidase plus CSL/SSL plus DATEM plus Ascorbic acid	4–5+ 6–8+ 6–7+ 8+ 800–100+ 600–800+ 55–60	Significant increase in volume and total score	Ma (2007)

4.15 MILK AND DRIED MILK POWDER

Milk and dried milk powder are often used in Guangdong-style steamed bread and buns to increase the flavor of the products and enhance nutritional quality. Milk and dried milk powder are added during the first dough mixing stage.

4.16 COLORINGS

Where approved by Food Law, food colorings can be added. An example is a combination allura red (129) and brilliant blue FCF (133), which gives an attractive purple taro color in some steamed breads and buns where no taro is included.

4.17 CONCLUSION

The addition of optional ingredients to steamed bread is only relatively recent. The number and variety of new products continues to increase and offers consumers greater choice and quality as well as the opportunity to introduce the much needed nutritional fortification. Product staling and short shelf life are issues that still need to be addressed in the future. The inclusion of microingredients, which have mainly been designed to function in Western bread-making systems, may provide some benefit, but may need modification or to be added in different combinations in steamed bread.

REFERENCES

Bao, Y.R., Wang, X.L., 2008. Effect of fungal α-amylase on the quality of steamed bread (Natural Science Edition) J. Henan Ind. Univ. 4, 35–37.

Bao, Q., 2010. Studies on the Effect of Particle Size of Bran to the Quality of Steamed Bread (M.Sc. thesis). Henan Industrial University, Zhengzhou, Henan, China (In Chinese, English abstract).

Bui, L.T.T., Small, D.M., Coad, R., 2013. The stability of water-soluble vitamins and issues in the fortification of foods. In: Preedy, V.R., Srirajaskanthan, R., Patel, V.B. (Eds.). Preedy, V.R., Srirajaskanthan, R., Patel, V.B. (Eds.), Handbook of Food Fortification and Health: From Concepts to Public Health Applications, vol. 1. Humana Press, NY, USA, pp. 199–211. http://dx.doi.org/10.1007/978-1-4614-7076-2.

Cao, Y., 2010. Preparation and Properties of Dietary Fibre From Barley Spent Grains and Its Application in Steamed Bread (M.Sc. thesis). Wuhan Industry College, Wuhan China (In Chinese, English abstract).

Cauvain, S.P., Huang, S., 1986. Chinese steamed bread. FMBRA Bull. 4, 151–158 (August).

Chang, Y.H., Liu, C.H., Huang, S.W., Zhang, X.K., 2010. Studies on effects of ascorbic acid on the quality of Chinese steamed bread. J. Grain Sci. Technol. Economy 35 (6), 43–44 (In Chinese).

Cruickshank, L.J., 1996. The Effect of Ingredients on the Quality of Northern Style Steamed Bread (B.Sc. thesis). University of New South Wales, Sydney, Australia.

Fu, L., Tian, J.C., Sun, Z., 2010. Effect of resistant starch on processing quality of northern style Chinese steamed bread. J. Chin. Cereals Oils Assoc. 25, 53–56 (In Chinese, English abstract).

Fu, Y.M., 2006. Studies on the Factors Affecting the Quality of Steamed Bread Made From the Mixture of Wheat Flour and Other Grain Flours (M.Sc. thesis). Henan University of Technology, Zhengzhou, China (In Chinese, English abstract).

Hao, M., Beta, T., 2012. Development of Chinese steamed bread enriched in bioactive compounds from barley hull and flaxseed hull extracts. Food Chem. 133, 1320–1325. http://dx.doi.org/10.1016/j.foodchem.2012.02.008.

He, C., Lin, X., Li, G., Yu, L., 2008. Research of steamed bread dough fermentation behavior. J. Food Res. Dev. 29 (9), 93–96 (In Chinese, English abstract).

Hu, L.S., Chen, J.R., Zhou, F., Li, Q.F., 2003. Study on the quality improvement of steamed bread flour using domestic lipase. Cereals Feed Ind. 2, 3–4 (In Chinese, English abstract).

Huang, S., Quail, K., 1997. Effects of ingredients on the quality of Guangdong style steamed bread. In: Tarr, A.R., Ross, A.S., Wrigley, C.W. (Eds.), Cereals 97 Proceedings of the 47th Australian Cereal Chemistry Conference. Cereal Chemistry Division, Royal Australian Chemical Institute, Melbourne, Australia, pp. 81–84.

Jiang, Z.Q., Cong, Q.Q., Van, Q.J., Kumar, N., Du, X.D., 2010. Characterisation of a thermostable xylanase from *Chaetomium* sp. and its application in Chinese steamed bread. Food Chem. 120, 457–462. http://dx.doi.org/10.1016/j.foodchem.2009.10.038.

Kruger, J.E., Morgan, B., Preston, K.R., Matsuo, R.R., 1992. Evaluation of some characteristics of Chinese steamed buns prepared from Canadian wheat flours. Can. J. Plant Sci. 369–375.

Lai, M.S., Lin, T.C., 2006. Bakery products. In: Hui, Y.H. (Ed.). Hui, Y.H. (Ed.), Handbook of Food Science, Technology and Engineering, vol. 4. CRC Press, Boca Raton, FL, USA (Chapter 148).

Li, L.T., Chen, X.Y., 1999. The improvement of steamed bread quality by the addition of lard. J. Chin. Cereals Oils Assoc. 14 (2), 47–50 (In Chinese, English abstract).

Limley, H.A., 2008. Wheat Quality Requirements for Char Siew Bao (BBQ Pork Steamed Buns) (Ph.D. thesis). Curtin University, Australia.

Lin, M.G., Ma, Y.S., Huang, S.W., He, Y.Q., Wang, H., 2010. Effect of cellulase on the characteristics of dough rheology and quality of steamed bread. Cereals Oils Process. 9, 56–58 (In Chinese).

Lin, S.Y., Chen, H.H., Lu, S., Wang, P.C., 2012. Effects of blending of wheat flour with barley flour on dough and steamed bread properties. J. Texture Stud. 43, 438–444. http://dx.doi.org/10.1111/j.1745-4603.2012.00352.x.

Lin, T.C., 1983. Effects of Processing Methods and Wheat Flours on the Quality of Steamed Bread (M.Sc. thesis) University of Kansas State, Kansas, USA (accessed 20.02.15.) http://hdl.handle.net/2097/12974.

Liu, Z.D., Dong, C.W., Pang, K., 2006. Techniques on Flour Quality Improvement and Application. Chinese Light Industry Publisher, Beijing, China (In Chinese).

Liu, C.H., 2005. Production Techniques of Steamed Products. Chemical Industry Publisher, Beijing, China (In Chinese).

Lu, Q., He, Y., Liu, X., 2015. Property assessment of steamed bread added with cellulase by using fuzzy mathematical model. J. Texture Stud. 46, 420–428. http://dx.doi.org/10.1111/jtxs.12141.

Ma, T., 2007. Application Studies of Enzymes, Emulsifiers and Ascorbic Acid to the Production of Steamed Bread (M.Sc. thesis). Henan Industrial University, Zhengzhou, Henan, China (In Chinese, English abstract).

Noor Aziah, A.A., Ho, L.H., Noor Shazliana, A.A., Bhat, R., 2012. Quality evaluation of steamed wheat bread substituted with green banana flour. Int. Food Res. J. 19, 869–876.

Polizeli, M.L.T.M., Rizzatti, A.C.S., Monti, R., Terenzi, H.F., Jorge, J.A., Amorim, D.S., 2005. Xylanases from fungi: properties and industrial applications. Appl. Microbiol. Biotechnol. 67, 577–591. http://dx.doi.org/10.1007/s00253-005-1904-7.

Pomeranz, Y., Huang, M.,L., Rubenthaler, G.L., 1991. Steamed bread III. Role of lipids. Cereal Chem. 68, 353–356.

Ren, H.T., Cheng, L.Y., 2007. Effect of combination addition of enzymes on the quality of steamed bread. J. Grain Oils Process. 9, 101–102 (In Chinese).

Sim, S.Y., Noor Aziah, A.A., Cheng, L.H., 2011. Characteristics of wheat dough and Chinese steamed bread added with sodium alginates or konjac glucomannan. Food Hydrocoll. 25, 951–957. http://dx.doi.org/10.1016/j.foodhyd.2010.09.009.

Su, D.M., 2005. Studies on Classification and Quality Evaluation of Staple Chinese Steamed Bread (Ph.D. thesis). Chinese Agricultural University, Beijing, China (In Chinese).

Wang, X.L., Wang, F.C., 2006. Effect of combining emulsifier and enzyme on the quality of steamed bread. J. Chin. Cereals Oils Assoc. 21 (3), 241–244 (In Chinese, English abstract).

Wang, S., Lin, K., Qin, X., Chen, X., Liu, P., Guo, X., Zhang, B., 2011. Effect of thermo-tolerant xylanase from thermophilic *Geobacillus* sp. PZHI on the quality and shelf-life of steamed bread. J. Food Sci. 32 (11), 137–141 (In Chinese, English abstract).

Wu, M.Y., Chang, Y.H., Shiau, S.Y., Chen, C.C., 2012. Rheology of fiber-enriched steamed bread: stress relaxation and texture profile analysis. J. Food Drug Anal. 20, 133–142. (accessed 02.06.15.) http://www.fda.gov.tw/en/publishJFDAList.aspx.

Xiao, A., He, D., Zhang, S., 2005. Promoting quality and postponing staling of steamed bread by adding baked soybean flour or dietary fiber from soybean hulls. J. Chin. Cereals Oils Assoc. 20 (4), 73–76 (In Chinese, English abstract).

Ying, B., Jun, L., Yunzi, F., Zhuo, C., Zaigui, L., 2009. Application of DATEM for Chinese steamed bread made from flours of different gluten qualities. J. Sci. Food Agr. 89, 227–231. http://dx.doi.org/10.1002/jsfa.3430.

Yu, K.P., 1990. Techniques Involved in the Art of the Southern Chinese Style Steamed Bread Making (M.Sc. thesis). University of New South Wales, Sydney, Australia.

Zhang, X., Sun, J., Li, Z., 2007. Effect of DATEM on dough rheological characteristics and qualities of CSB and bread. Cereal Chem. 84, 181–185.

Zhang, J., Zhao, Y., Cheng, H.H., Zhang, Y.Y., Wang, Y.S., 2014. Effects of ascorbic acid combined with endoxylanases on the shape and quality of steamed bread. J. Grains Edible Oils 4, 39–42 (In Chinese, English abstract).

Zheng, X.L., Ma, L., Li, L.M., 2006. Effect of de-fatted soy flour on the quality of wheat flour and steamed bread. Grains Feed Ind. 4, 15–16 (In Chinese, English abstract).

Zhu, F., 2014. Influence of ingredients and chemical components on the quality of Chinese steamed bread. Food Chem. 163, 154–162. http://dx.doi.org/10.1016/j.foodchem.2014.04.067 0308.

Chapter 5

Steamed Buns and Steamed Rolls—Fillings and Their Preparation

5.1 INTRODUCTION

Steamed buns and rolls are more popular than steamed bread in southern China and Southeast Asia. These products, often consumed as snacks, must have an attractive appearance and appealing flavor and so have evolved to suit local preferences. An example is Malaysian *pau*, which may be filled with chicken curry or beef curry. The names of steamed buns are often derived from the name of their fillings, such as *char siew bao*, in which *char siew* (Chinese barbecue pork) is the basis of the filling.

Filling may account for approximately 40–50% of the volume of the steamed bun (Anon., 1989), whereas the filling for steamed rolls may only be thinly spread between the dough layers. Needless to say high-quality fillings are essential for high-quality buns.

5.2 SAVORY FILLINGS

5.2.1 Introduction

Savory fillings may contain both raw and cooked ingredients. Fillings may contain vegetables, meat, or seafood either alone or in combination. Seasonings and flavorings are added to suit local preferences.

5.2.2 Vegetable Fillings

5.2.2.1 Fresh Vegetable Fillings

Examples of fresh vegetables are given in Table 5.1. Fresh vegetables may be lightly cooked, steamed, or salted. Lightly cooked leafy and green vegetables are refreshed immediately in cold water to stop the cooking process and maintain their green color. Cooked vegetables may be finely cut, processed, or drained before use.

Steamed Breads. http://dx.doi.org/10.1016/B978-0-08-100715-0.00005-7

TABLE 5.1 Ingredients for Savory Fillings Used in Steamed Buns and Steamed Rolls

Type	Examples
Fresh vegetables	Green leafy vegetables, Chinese celery, turnip, cabbage, Chinese cabbage, spinach, winter bamboo shoot, shepherd's purse (*Capsella bursa-pastoris*, a member of the mustard family), pumpkin, leeks, chives
Dried and preserved vegetables	Dried daylily, Chinese mushrooms, edible fungus, dried winter bamboo shoots, preserved vegetables such as *meigancai* (mustard greens and cabbage)
Meats	Pork, beef, chicken, lamb, *char siew pork*, pork sausage (*lup cheong*), barbeque pork
Seafood	Prawns, crabs, crab roe, scallops, crab sticks, fish cake
Other ingredients	Fats and oils, bean curd—plain or flavored (often used in vegetable fillings), dried starch noodles, *fun pi* (starch sheet jelly), fried gluten balls, egg, cornstarch
Seasonings	Ginger, shallots, Chinese cooking wine (*Shaohsing*), pepper, salt, sugar, chicken powder, five spice powder, Sichuan peppercorn powder, soy sauce, oyster sauce, sesame oil, monosodium glutamate, curry paste, chili, coriander, garlic, dried shrimps

Fat, oil, starch, or raw egg can be added to optimize the texture and stickiness of the mixture before the addition of seasonings and flavorings.

5.2.2.2 Dried Vegetable Fillings

Chinese dried vegetables such as dried daylily, tips of bamboo shoots, Chinese mushrooms, and edible fungus are the main ingredients used for this type of filling (Table 5.1). Starch noodles can be added to create texture. Dried vegetables should be soaked in warm water to soften before use. They are then cut into small pieces and mixed with any seasonings, flavorings, or thickeners during cooking (Anon., 1989).

5.2.3 Meat Fillings

Pork is the most popular meat filling but beef, lamb, and chicken are also used. Cuts should contain both lean meat and fat, to give tender eating quality. Pork is not used in products needing Halal certification.

5.2.3.1 Raw Meat Fillings

Raw meat fillings are widely used. Pork, chicken, beef, and lamb can be used alone, or combined with seafood and/or vegetables. Raw meat fillings should be tasty, tender, and juicy.

In southern China, sugar is often added to the fillings. Water or broth or a broth-based jelly may also be added to achieve the required consistency and water-holding capacity of the fillings (Huang, 1999).

5.2.3.2 *Cooked Meat Fillings*

There are two methods used to prepare cooked meat fillings.

The first method uses raw meat cooked with the other ingredients. For example, curried meat fillings, popular in Southeast Asia, are prepared by frying curry paste, either handmade or a commercial preparation, with the meat and other ingredients. The filling should be cooled before use.

The second method uses precooked meat cooked with the other ingredients. *Char siew* or barbequed pork is a popular example. Typically, the chopped barbecued pork is stir fried with onions before soy sauce, sugar, and other ingredients are added.

5.2.4 Mixture of Vegetable and Meat Fillings

This type of filling is made by mixing processed raw or cooked vegetables and meat (Huang, 1999). The most popular fillings are a mixture of chopped pork and fresh vegetables (*cai rou xian*) and a mixture of minced meat and preserved vegetable (*meigancai rou xian* and *dongcai rou xian*).

5.3 SWEET FILLINGS

Sweet fillings are popular, particularly in southern China and other parts of Asia. The fillings are prepared by mixing sugar with various other ingredients such as seeds, tubers, fruits, and eggs. Examples of the different types of sweet fillings are given in Table 5.2.

Legumes such as adzuki and mung beans need to be soaked before boiling. Water is then added and the well-cooked beans are mashed to enable the seed

TABLE 5.2 Ingredients for Sweet Fillings Used in Steamed Buns and Steamed Rolls

Type	Examples
Sweet pastes	Adzuki (red or black bean), mung bean, taro, sweet potato, yam, lotus seed, sesame seed, finely chopped dates, nuts, preserved or dried fruits, chocolate
Sugar filling	White sugar, white/brown sugar, palm sugar, coconut sugar
Egg custard	Egg yolk, corn starch, milk, cream, sugar
Flavorings	Lard, sesame oil, sesame seed, preserved rose flower, osmanthus flowers, pandan, cocoa powder

coat to be strained out. Excess water in the mash is removed by filtration before sweetening with sugar. Seeds, fruits, or tubers are cooked with sugar and other minor ingredients before grinding to a fine paste. Pastes such as taro are stir fried with oil to reduce the moisture content. Cooked fillings need to be cooled before use.

Manufacture of steamed buns and steamed rolls and details of how the fillings are used will be discussed in Chapter 7.

5.4 CONCLUSION

Fillings for steamed buns provide variety and result in a large range of products which are available to the consumer. There are also a range of regional preferences for ingredients content and their preparation.

REFERENCES

Anon., 1989. Techniques for Making Wheaten Foods, eleventh ed. Chinese Commercial Publishing House, Beijing, China (In Chinese).

Huang, S., 1999. Wheat products: 2. Breads, cakes, cookies, pastries, and dumplings. In: Catharina, Y.W., Liu, K.S., Huang, Y.W. (Eds.), Asian Foods Science & Technology, first ed. Technomic Publishing Co Inc., Lancaster, PA, USA, pp. 71–108.

Chapter 6

Dough Preparation and Processing

6.1 INTRODUCTION

As with Western bread, the commercial process for making steamed bread consists of a number of basic unit operations. These can be summarized as follows:

1. Selection of ingredients according to the product being manufactured
2. Weighing or otherwise accurately dispensing the ingredients
3. Mixing the dough
4. Fermentation—a number of different methods are used
5. Molding
6. Proofing at carefully controlled temperature and humidity
7. Steaming
8. Cooling and packing

Fig. 6.1 shows the flow diagram for two typical processing procedures used for steamed bread and bun production. These stages will be explained over the next two chapters. In this chapter, dough processing will be discussed up to the fermentation stage, with the remaining steps covered in the following chapter.

6.2 SELECTION AND PREPARATION OF INGREDIENTS

The essential dough ingredients of flour, water, and yeast or sourdough are discussed in Chapter 3, and the optional ingredients which add variety are covered in Chapter 4. Special care should be taken to ensure that the quality of the major ingredient—flour—is suitable for the product being manufactured. All ingredients should be assessed for quality and adherence to specification on receipt from the supplier.

Ingredients are prepared and set aside according to the formula for each production batch. Small or trace amounts of ingredients may need to be dissolved or suspended in water or diluted with flour or starch as a premix. Flour and water temperature should be monitored to ensure the final dough temperature is within the required range.

Steamed Breads. http://dx.doi.org/10.1016/B978-0-08-100715-0.00006-9

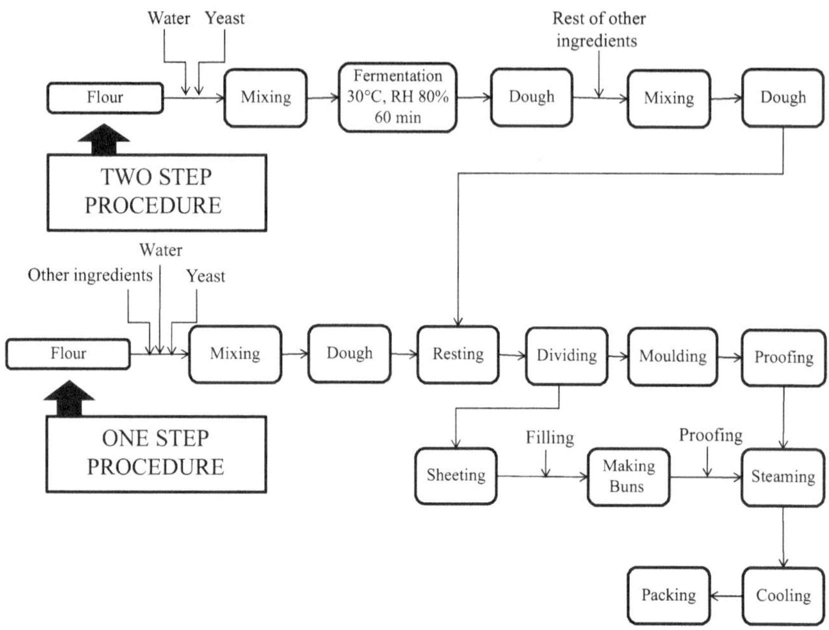

FIGURE 6.1 Flow diagram of the one-step and two-step procedures used for making steamed bread and buns. *Reproduced with permission from Huang, S., 2014. Steamed bread. In: Zhou, W., Hui, Y.H., (Eds.) Bakery Products Science and Technology, second ed. Wiley Blackwell, Chichester, West Sussex, UK, pp. 539–562.*

The order in which ingredients are added to the mixer may affect the quality of the end product. For instance, in the two-step procedure, water, sugar, and improver are mixed separately. Instant active dry yeast and flour are dry blended before water is added. This ensures yeast will not be inhibited by direct contact with very cold or very hot water. Fat addition should be delayed until the dough is partially mixed. Salt addition is often delayed until 5–6 min before the end of mixing to improve dough development.

6.3 DOUGH MIXING

Control of the mixing procedure is important for the final quality of steamed bread. The aims of mixing are to:

1. Mix and hydrate the ingredients
2. Develop dough to the required consistency and temperature

During mixing, the ingredients are transformed to a homogenous mass as the ingredients hydrate and the gluten network forms. As the dough develops, air is incorporated in a network of gas cells and the dough is able to withstand further processing.

6.3.1 Types of Mixers

Mixers are commonly described as low, medium, and high intensity. The most popular ones used for steamed bread manufacture are the medium-intensity mixers which include spiral, planetary, and horizontal types. Horizontal drum mixers with a capacity of 50–200kg are widely used in larger factories in China. Mixing time is up to 15min, which gives some tolerance if delays occur later in the process. Finished dough temperature must be monitored and ingredient water temperature controlled. High-intensity mixers can achieve optimum dough development within 4–5min, but finished dough temperatures can be higher than desirable due to the high work input; therefore a high level of process control is required to maintain dough temperature. A Tweedy high speed mixer fitted with a water jacket was successfully used by Cauvain and Huang (1986) to apply the Chorleywood Bread Process to steamed bread production.

6.3.2 Bakery Water Addition

In Western bread production, about 55–65% water is added to the dough, based on flour weight. Chinese steamed bread contains 10–15% less added water (Huang and Moss, 1991). The lower water addition results in a characteristic chewy texture (Faridi and Rubenthaler, 1983; Lukow et al., 1990).

Maximum water addition for northern-style steamed breads can be up to 45% by flour weight (Huang and Miskelly, 1991; Su, 2005), although the rate used in industrial production is slightly lower at 36–42% (Liu, 2005). Maximum water addition for southern-style steamed breads and buns as well as Guangdong-style steamed bread can be as high as 50% by flour weight (Shuai, 1998).

Appropriate bakery water addition can be estimated on the basis of farinograph water absorption. For northern style, this is about 70% of farinograph water absorption (Huang et al., 1993; Su, 2005), and for southern style about 80% (Huang et al., 1998; Su, 2005).

These low water additions result in doughs that would be considered "tight" in bakeries producing Western bread, but is essential for subsequent sheeting, and to give the preferred shape profile and texture in the finished product.

6.3.3 Dough Microstructure

Reduced water addition to steamed bread doughs affects the mechanism of gluten development during mixing. Unlike Western bread, the microstructural appearance of steamed bread doughs shows little change during the mixing process. In Western bread, gluten becomes hydrated and forms a cohesive mass as gluten strands form and become finer and interconnected. In steamed bread, there is little initial pulling away of the gluten from the starch and the gluten continues to surround the starch as mixing progresses. Nonetheless, adequate mixing time is still required for good steamed bread quality. The mechanism of dough development is thought to be similar to that for noodle doughs (about

32% water addition), where mixing serves to hydrate the dough and further development occurs during sheeting (Huang and Moss, 1991). In the case of steamed bread, further gluten development and realignment would be expected during subsequent sheeting steps.

6.3.4 Mixing Variables

6.3.4.1 Mixing Time

Optimization of dough mixing is important for subsequent dough handling and product quality. Mixing time will vary according to the type of mixer and mixing speed, flour properties, water addition, and temperature. Flour variables affecting mixing time include strength and water absorption. Other ingredients such as fat, salt, enzymes, and improvers also affect mixing time. Fig. 6.2 shows the same dough at two different mixing times.

6.3.4.2 Finished Dough Temperature

Finished dough temperature needs to be controlled as it affects the fermentation rate and subsequent processing. It is mainly affected by the temperature of the added water, plus temperature of flour and other ingredients. The room temperature (if uncontrolled) and amount of heat generated during the mixing process are also key variables. By adding cooler or warmer water at the mixing

FIGURE 6.2 Steamed bread dough after 1 min mixing (left) and fully developed after 12 min mixing (right).

stage it is possible to achieve the correct dough temperature. This will vary according to the type of product being manufactured, but about 28°C is ideal finished dough temperature. In tropical areas, ice water may need to be used when the ambient temperature is above 35°C. Larger production facilities may employ jacketed mixers where jacket water temperature is used to maintain correct dough temperature.

Selecting the correct dough water temperature is largely based on experience, but can be estimated in the following manner. First, what is sometimes called a "friction factor"—the temperature increase of the dough during mixing in a particular mixer—is calculated during production.

$$F = 3 \times T_{fd} - (T_{room} + T_{flour} + T_{water}) \tag{6.1}$$

where F = friction factor, T_{fd} = finished dough temperature, T_{room} = room temperature, T_{flour} = flour temperature, T_{water} = water temperature.

Second, the temperature of the water needed to achieve a desired finished dough temperature can then be estimated by the following formula:

$$T_{water} = 3 \times T_{fd} - (T_{room} + T_{flour} + F) \tag{6.2}$$

As an example, if F = 4°C as calculated from Eq. (6.1), then using Eq. (6.2) where T_{room} = 24°C, T_{flour} = 23°C, and T_{fd} = 28°C, then T_{water} added to the dough needs to be 33°C.

6.4 FERMENTATION

Fermentation is a critical step in steamed bread production. During fermentation, added yeast and bacteria (mostly *Lactobacillus* spp.) naturally present in the flour or starter doughs use enzymes to convert sugars to carbon dioxide, alcohol, and heat. They also provide flavor derived from fermentation by-products.

6.4.1 Function of Yeast and Other Microflora

The most apparent outward change as carbon dioxide is released by the yeast during fermentation is a steady increase in the volume of dough mass. The gluten structure is developed as the gas bubbles expand and by various factors such as proteolytic enzymes, alcohol, and the various organic acids formed during fermentation.

Fermentation is affected by the quantity and quality of yeast, dough temperature, available fermentable ingredients and salt. As the concentration of yeast is increased, gas production is increased. As the dough temperature increases, gas production is increased, reaching an optimum at about 38–40°C, after which thermal deactivation of the yeast begins. Yeast deactivation is considered complete at about 60°C.

Yeast, like any other living organism, requires nutrients such as an adequate supply of fermentable carbohydrates and available nitrogenous substances, as well as certain essential minerals. Correct water addition is important to activate the yeast. Yeast addition needs to be adjusted according to different styles dough preparation methods and recipes. (See Chapter 3 for the different types of yeast and their usage rates.)

6.4.2 Dough Maturity

As fermentation progresses, changes occur in the physical character of the dough; these are described as dough ripening or mellowing. The character of a "mellow" gluten is even more important to steamed bread than for Western bread in retaining a large volume and porous structure. Steamed bread is more sensitive to outside stresses because it does not have a solid crust. The high rate of collapse of steamed bread made from undermixed and underfermented dough confirms this importance. This will be discussed further in Chapter 10.

A skilled operator is required to properly judge the maturity of steamed bread dough during fermentation. In the case of the sourdough procedure described later in this chapter, dough has reached the right maturity when it has increased in bulk, looks white and smooth, is soft to the touch, and has a normal alcohol and sour smell that is not irritating. There are a number of qualitative methods used to judge dough maturity. For example, a small indentation made in the dough with a finger will spring back the original shape and dough is extensible when a small piece is pulled from the bulk. Cut dough will have even gas cells, about the size of a soybean. Overfermented dough is very soft, and will have an irritating sour smell.

6.5 DOUGH PREPARATION PROCEDURES

There are three major and two minor dough preparation procedures (Fig. 6.3). These are:

- One-step procedure
- Two-step procedure
- Sourdough procedure
- Liquid ferment procedure
- Fermented glutinous rice (*tiang jiu niang*) procedure

The liquid ferment and fermented glutinous rice procedures are not used routinely in commercial production. They are typically used in regional or home manufacture as there can be some negative aspects of their use. Steamed bread made by the fermented glutinous rice procedure is a traditional food in China, and is very popular in Jinhua region, Zhejiang province.

6.5.1 One-Step Procedure

The one-step procedure (Figs. 6.1 and 6.3) which has also been described as the "no time" procedure is favored by small to medium manufacturers, as it

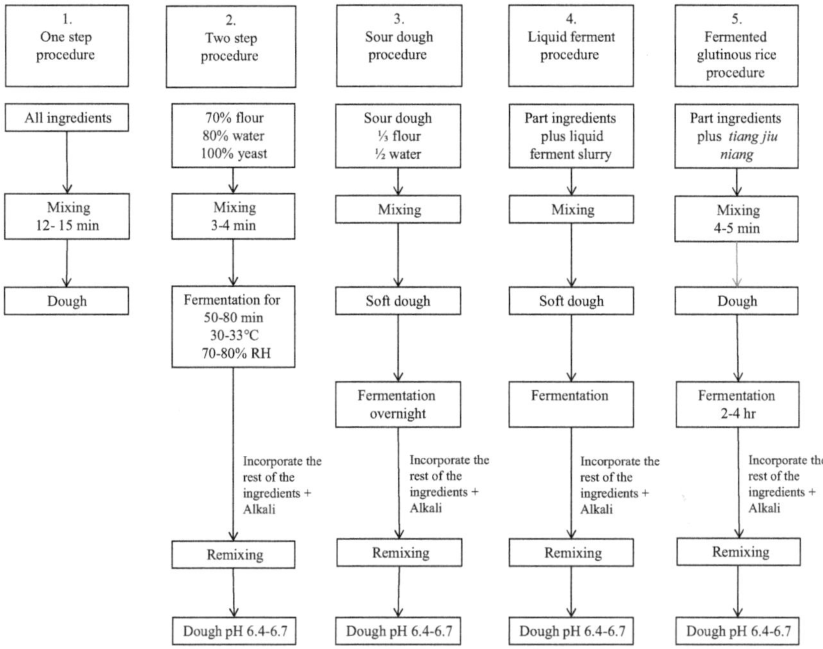

FIGURE 6.3 Five dough preparation procedures for steamed breads and buns.

takes the least time. In its basic form a lean formulation is used with 36–42% water and 0.3–0.4% dried yeast or 1% compressed fresh yeast. Fat and sugar can also be added for Guangdong-style products. Sourdough cannot be used in this procedure because of the relatively short fermentation time. The ingredients are mixed together for 12–15 min in a medium-intensity mixer. Final dough temperature should be 33–35°C in winter and 30–32°C in summer (Liu, 2005). If necessary, ingredient water temperature is used to control the dough temperature. Unlike Western bread there is only a very short intermediate fermentation step. Even 20-min intermediate fermentation can result in poorer spread ratios, inferior crumb texture, and lack of skin smoothness and shininess (Kim et al., 2001). After mixing, the dough is divided, molded, and proofed. The final proofing is usually carried out at 38–40°C and 85–90% relative humidity (RH) for about 60 min (Liu, 2005). Products made by the one-step procedure are the least preferred by consumers because they have less flavor and stale quickly. However, the process has many advantages, especially in terms of improved production efficiency (Table 6.1).

6.5.2 Two-Step Procedure

The two-step procedure (Figs. 6.1 and 6.3) is similar to the sponge and dough method in Western bread making in that a proportion of flour, yeast, water, and

TABLE 6.1 Advantages and Disadvantages of the Different Production Methods for Steamed Bread Production

Method	Advantages	Disadvantages
One-step procedure	Reduced fermentation losses, processing time, labor, power, and space and equipment requirements Improved production efficiency No need for dough neutralization Easier to mold	Less flavor and aroma compared with longer fermentation methods Products have finer texture Products stale more quickly Yeast usage is higher than in the two-step procedure Mixing regimen needs to be strictly controlled
Two-step procedure	50% lower yeast requirement compared with one-step method Larger volume Smoother appearance Better texture Improved flavor Lower rate of substandard products Better keeping quality Better flexibility and process control	Increased labor cost, power consumption, and wear and tear on equipment Higher fermentation losses Lots of dough on hand which can spoil in case of power or equipment failure Longer processing time
Sourdough procedure	Low cost and availability of sourdough Variety of fermentation dough troughs can be used Can be used without temperature and humidity control in the bakery Strict process control not required Unique flavor and texture of products Good shelf life	Dough neutralization a skilled operation Poor process control Risk of uncontrolled microbiological conditions
Liquid ferment procedure	Low cost and availability of sourdough Fermentation is easily managed Good steamed bread flavor and texture	Control of pH difficult Neutralization of the dough requires a lot of skill Risk of toxic bacteria Risk of uncontrolled microbiological conditions
Fermented glutinous rice procedure	Very good quality and flavor Low cost	Not suitable for commercial production

other ingredients is mixed into a "sponge" and left to ferment before remixing with the remaining ingredients. In the first mixing phase, 70% of the flour, all of the yeast, and 80% of the total water are mixed together for 3–4 min, with a finished dough temperature of 28–30°C (Liu, 2005). Doughs are fermented at 30–33°C and 70–80% RH for 40–90 min. Fermentation time can be adjusted by varying the yeast addition as well as temperature and humidity (Liu, 2005). Fermentation time is less than that used in Western sponge and dough bread production where sponge times of 2.5–6 h are frequently encountered in commercial practice (Pyler, 1982).

In the "dough" stage, the remaining flour and water are added, plus alkali solution and other ingredients such as sugar and salt. The dough is remixed for about 8–12 min to optimum dough development. The final dough temperature is about 33–35°C and the pH between 6.2 and 6.7. The dough is molded by hand or machine before proofing for 60 min at 38–40°C and 80–90% RH.

The two-step procedure is used in medium-sized semi-automated or fully automated production lines, where the process can be controlled. Steamed breads made by this process have better texture, flavor, and keeping quality than breads made by the one-step process (Table 6.1).

6.5.3 Sourdough Procedure

There is increased interest in steamed bread made by the sourdough procedure because of better flavor and the perceived health benefits (Wu et al., 2012; Zhu, 2014). There are a number of methods of production. Generally, a starter dough is prepared from the dough of a previous batch. The left over dough is mixed with flour and water and allowed to ferment overnight at ambient conditions for use in the next day's production. In Thailand, preparation of a starter dough can take several days and varies according to region (Keeratipibul and Luangsakul, 2012). If a commercially available starter is used, it needs to be soaked in water before use.

In the first dough mixing stage shown in Fig. 6.3, 5–15% of sourdough is mixed with 33% of the total flour and about half of the water (the total water used is about 37–42% by flour weight). Instant dried yeast may be used as a supplement. The dough is mixed until fully developed. Fermentation is often carried out under ambient conditions, which can be up to 10 h in a Chinese winter. Where there is no temperature control in the proofing area, a thermostated electrically heated cable (heat trace) can be wrapped around the dough trough. After fermentation, the dough becomes sour because of the organic acids produced by *Lactobacillus* spp. The fermented dough is transferred to the mixer and the remaining flour and water are added.

Because the dough pH has fallen, it needs to be neutralized with alkali back to pH 6.2–6.7. If the dough is underneutralized, the steamed bread has a grayish color, small volume, dense texture, and sour smell and may collapse on reheating. If the dough is overneutralized, both the dough and the steamed bread will have yellow color with a strong alkali smell. The correct level of alkali addition can be determined by several methods including assessing dough smell (Anon 1989). A skilled operator can judge the degree of neutralization of the dough by smelling the balance between the sour notes and the alkali notes. The dough needs to be carefully remixed to ensure even dispersion of the alkali.

6.5.4 Liquid Ferment Procedure

6.5.4.1 Liquid Ferment Slurry

Residual sourdough is used to make a liquid ferment for steamed bread production. The sourdough is soaked and dispersed in warm water before being mixed into a slurry with flour and yeast. The slurry is then fermented for 3–4 h at 25–30°C before use.

6.5.4.2 Liquid Ferment Dough Preparation

The liquid ferment slurry is then transferred to a mixer with sufficient flour and water to make a dough. Alkali solution is added after fermentation as for the sourdough procedure. Remixing, dividing, molding, proofing, and steaming are all similar to the sourdough procedure.

The advantages of the liquid fermentation procedure are:

1. Low cost; sourdough is easily available and cheap.
2. Management of the ferment is relatively easy.
3. The flavor and texture of the final products are as good as that of sourdough fermentation.

The disadvantages of the system are:

1. Control of the correct pH of the final dough is difficult.
2. It is possible to get growth of undesirable or possibly toxic bacteria in the liquid ferment if pH and temperature are not strictly controlled.

6.5.5 Fermented Glutinous Rice (*Tiang Jiu Niang*) Procedure

Fermented sweet rice (*tiang jiu niang)* is a popular food in China. It is produced from steamed and cooled glutinous rice to which small quantities of wheat flour and *jiu zao* (spent distillers grains) are added. The ingredients are mixed and allowed to ferment. The product is ready for consumption after 48 h. *Jiu zao* contains molds such as *Rhizopus, Mucor, Monilia*, and *Aspergillus* (Wang, 1991). Some yeast and bacteria may also be involved in the fermentation process (Luh, 1999). *Tiang jiu niang* contains 1.5–2.0% alcohol with an acidity of 0.5–0.6% as lactic acid and some glucose, maltose, and oligosaccharides.

Tiang jiu niang is consumed largely in winter and is also used for steamed bread fermentation where it provides a good and cheap source of yeast and bacteria. The quality and flavor of steamed bread made by this method are very good. This procedure may also be known as the rice sauce or rice wine fermentation procedure because further fermentation of the mixture will produce either rice wine or rice vinegar.

For example, 20 kg of flour and 0–20 g instant dry yeast are given an initial brief mix, and then 8 kg of *tiang jiu niang* and 3 kg of water are added gradually and mixed for 4–5 min. The dough is then fermented for 2–4 h at 30–35°C and 70–80% RH. The fermented dough is then remixed with 5 kg flour, 50–100 g of alkali, and about 2 kg water (depending on the softness of the dough) for 7–10 min to optimum dough development. The final stages of the procedure are the same as for the sourdough procedure (Liu, 2005).

Although this procedure is cheap and produces a superior flavor in the steamed bread, it is not easy to control the fermentation. Therefore, it is not suitable for semi-automated or fully automated production of steamed products. Some small steamed bread factories and households in southern China may still use this procedure.

Advantages and disadvantages of the various dough preparation procedures are compared in Table 6.1.

6.6 CONCLUSION

The one-step, two-step, and sourdough procedures are the most important methods of dough preparation for steamed bread. The first two are the most suitable for application in large-scale industrial production as the process can be readily controlled. The sourdough procedure gives products with good flavor and texture, but it is more difficult to control the process.

REFERENCES

Anon, 1989. Techniques for Making Wheaten Foods, eleventh ed. Chinese Commercial Publishing House, Beijing, China (In Chinese).

Cauvain, S.P., Huang, S., 1986. Chinese steamed bread. FMBRA Bull. No. 4 151–158 August.

Faridi, H.A., Rubenthaler, G.L., 1983. Laboratory method for producing Chinese steamed bread and effects of formula, steaming and storage on bread starch gelatinization and freshness. In: Proceedings 6th International Wheat Genetics Symposium, Kyoto, Japan, pp. 863–867.

Huang, S., 2014. Steamed bread. In: Zhou, W., Hui, Y.H. (Eds.), In Bakery Products Science and Technology, second ed. Wiley Blackwell, Chichester, West Sussex, UK, pp. 539–562.

Huang, S., Miskelly, D., 1991. Steamed bread – a popular food in China. Food Aust. 43, 346–351.

Huang, S., Moss, R., 1991. Light microscopy observations on the mechanism of dough development in Chinese steamed bread production. Food Struct. 10, 289–293.

Huang, S., Betker, S., Quail, K., Moss, R., 1993. An optimised processing procedure by response surface methodology (RSM) for northern-style Chinese steamed bread. J. Cereal Sci. 18, 89–102. http://dx.doi.org/10.1006/jcrs.1993.1037.

Huang, S., Quail, K., Moss, R., 1998. The optimisation of a laboratory processing procedure for southern style Chinese steamed bread. Int. J. Food Sci. Tech. 33, 345–357. http://dx.doi.org/10.1046/j.1365-2621.1998.00166.

Keeratipibul, S., Luangsakul, N., 2012. Chinese steamed buns. In: Hui, Y.H., Özgül Evranuz, E. (Eds.), Handbook of Plant-Based Fermented Food and Beverage Technology. CRC Press, FL, USA, pp. 543–556.

Kim, C.S., Hwang, C.M., Song, Y.S., Kim, H.I., Chung, D.J., Han, J.H., 2001. Commercial wheat flour quality and bread making conditions for Korean-style steamed bread. J. Korean Soc. Food Nutr. 30, 1120–1128 (In Korean, English abstract).

Liu, C.H., 2005. Production Techniques of Steamed Products. Chemical Industry Publisher, Beijing, China (In Chinese).

Luh, B.S., 1999. Rice products. In: Catharina, Y.W., Liu, K.S., Huang, Y.W. (Eds.), Asian Foods Science & Technology, first ed. Technomic Publishing Co Inc., Lancaster, PA, USA, pp. 5–42.

Lukow, O.M., Zhang, H., Czarnecki, E., 1990. Milling, rheological and end-use quality of Chinese and Canadian spring wheat cultivars. Cereal Chem. 67, 170–176.

Pyler, E.J., 1982. Baking Science and Technology, third ed. Siebel Publishing Company, Chicago, ILL, USA (Chapter 15).

Shuai, K., 1998. Selection of Guangdong Dim Sum, second ed. Guangdong Science and Technology Publisher, Guangzhou, China (In Chinese).

Su, D.M., 2005. Studies on Classification and Quality Evaluation of Staple Chinese Steamed Bread (Ph.D. thesis). Chinese Agricultural University, Beijing, China.

Wang, H.H., 1991. Fermented rice products. In: Luh, B.S. (Ed.), Rice Utilization, vol. II. second ed. Van Nostrand, Reinhold, New York, pp. 195–223.

Wu, C., Liu, R.S., Huang, W.N., Rayas-Duarte, P., Wang, F., Vao, V., 2012. Effect of sourdough fermentation on the quality of Chinese northern-style steamed bread. J. Cereal Sci. 56, 127–133. http://dx.doi.org/10.1016/j.jcs.2012.03.007.

Zhu, F., 2014. Influence of ingredients and chemical components on the quality of Chinese steamed bread. Food Chem. 163, 154–162. http://dx.doi.org/10.1016/j.foodchem.2014.04.067 0308.

Chapter 7

Processing—Making Products

7.1 INTRODUCTION

Dough preparation methods are described in Chapter 6. This chapter will discuss the further processing of dough into finished products by sheeting, molding, dividing, proofing, steaming, cooling, and packing.

7.2 MOLDING

Dough molding is very important to the quality of the finished steamed products. Correct molding will ensure a fine and even crumb structure. A number of molding techniques for steamed bread, buns, and rolls are possible using hand or machine methods.

7.2.1 Steamed Bread Molding

7.2.1.1 Molding by Hand

For steamed bread made at home the dough is divided into the required size (about 140 g for traditional northern-style steamed bread) and the pieces rounded by hand. Dough can also be rolled into a long cylinder and cut to the required length for individual pieces by hand using the same technique as is used traditionally in small manufacturing facilities (Fig. 7.1). Hand molding is traditional for most plain and fancy homemade steamed breads, buns, and rolls. Even at present, some commercial high-quality steamed rolls and most *dim sum* items use a hand molding step as part of an otherwise mechanized process.

7.2.1.2 Molding by Machine

Machine molding enables production of large volumes of products of uniform size, shape, and appearance with a fine and even crumb structure. Most mechanical molders incorporate sheeting rolls, which will result in a reduction of the number and size of gas bubbles, and in turn reduce surface faults such as blisters and dimples. Further dough development occurs during sheeting, leading to an improvement in underdeveloped doughs and dough color.

Steamed Breads. http://dx.doi.org/10.1016/B978-0-08-100715-0.00007-0

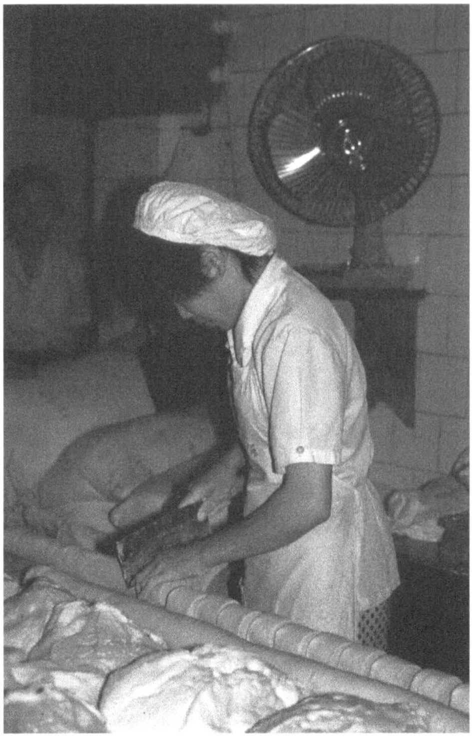

FIGURE 7.1 Steamed bread molding by hand.

7.2.1.3 Round Shape Molding

Round molding machines (*mantouji*) have been used widely in northern China for many years for steamed bread. Fig. 7.2 shows an early model which is typically used in small steamed bread factories. The molding machine consists of a dough hopper, divider, and molding chamber containing a pair of screw shaping rolls (Fig. 7.3). Dough is placed into the hopper, from where it is extruded and cut to size. The cut dough pieces are caught in the first cavity of the screw rolls and are formed into a round shape as the screw turns and conveys them along its length (Figs. 7.2 and 7.3). For further information on the operation of round molding machines see https://www.youtube.com/watch?v=vpaNiSln23E.

7.2.1.4 Rectangular Shape Molding

Mechanized molding machines begin forming the dough through a series of laminating and/or sheeting rolls. The formed dough sheet is then slit to the required width, rolled into a cylindrical shape, and finally cut into individual pieces (Fig. 7.4).

FIGURE 7.2 Steamed bread round molding machine. *Reproduced with permission from Huang, S., Miskelly, D., 1991. Steamed bread – a popular food in China. Food Aust. 43, 346–351.*

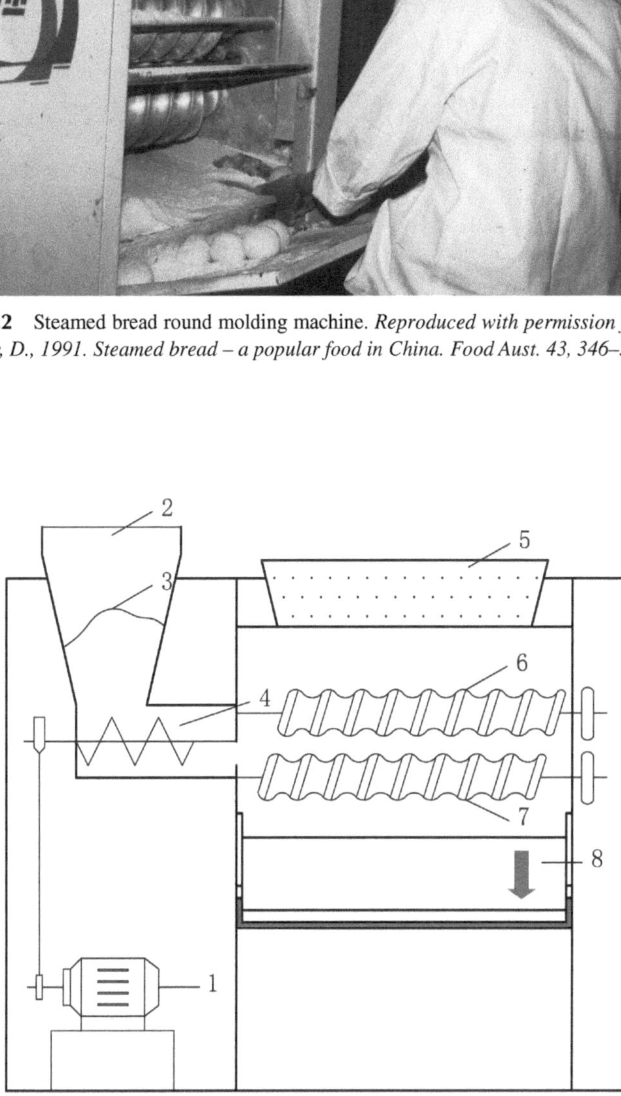

FIGURE 7.3 Diagram of horizontal steamed bread molding machine. (1) motor; (2) dough hopper; (3) dough; (4) extrusion chamber; (5) dusting flour; (6) upper shaping roll; (7) lower shaping roll; (8) exit chute. *Courtesy of Miss Chengyi Zhao.*

FIGURE 7.4 Machine-formed square-shaped *mantou. Courtesy Yang Jenq Machinery (Wuxi) Co. Ltd.*

As well as traditional *mantou*, the system is also able to produce filled products by extruding a filling into the center of the dough sheet as it is rolled into a cylinder along a conveyer belt (Fig. 7.7). For machine operation see http://baidu.ku6.com/watch/4796777567188854067.html?&recFrom=site&list=1

7.2.2 Steamed Bun Molding

7.2.2.1 Molding by Hand

The dough is rolled into a 4-cm-wide cylinder and individual pieces are cut from the roll similar to that shown in Fig. 7.1. The pieces are turned onto the cut end and flattened with the palm of the hand. They are then rolled to a circle using a rolling pin and rotated to create thinner edges (Fig. 7.5A). A small amount of filling is placed on the dough circle (Fig. 7.5B), and the edges pinched and pleated together (Fig. 7.5C). The finished bun is shown in Fig. 7.5D.

7.2.2.2 Molding by Machine

Equipment suited to automated manufacture of steamed breads is becoming popular in medium to large production facilities. Typical installations will include multiple dough sheeters, sheet dividers, filling depositors, and molding, forming, and sealing machines.

There are basically two approaches to mechanized production. The first approach is known as encrusting and the second approach is known as sheeting, filling, and forming. This latter technique has the advantage of producing

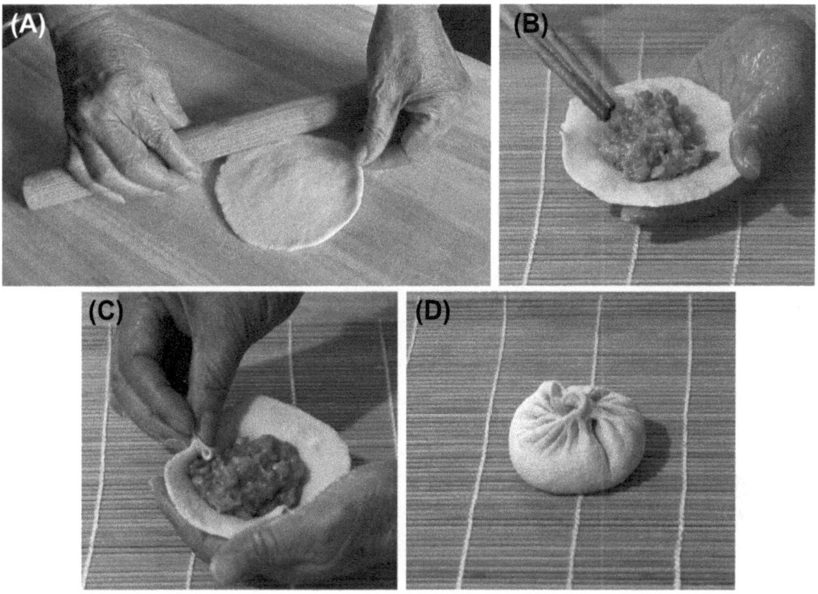

FIGURE 7.5 Hand molding steamed buns: (A) rolling, (B) filling, (C) folding, (D) finished bun.

a finer and more even crumb in the finished bun. Machine manufacturers can supply a range of forming attachments that enable manufacture of many different shaped and sized products. Examples of fillings include meat, vegetable, red bean paste, and *char siew* as detailed in Chapter 5.

7.2.2.3 Encrusting and Forming

The oldest technique for machine bun molding is referred to as *encrusting*. Encrusting is a process where the dough and filling are initially prepared and extruded separately before merging in a coextrusion step to enclose the filling in the dough. The dough tube and enclosed filling then passes through a forming and cutting shutter die to produce the finished bun. An example of an encrusting machine is shown in Fig. 7.6. There are currently several manufacturers of encrusting machines capable of producing steamed buns as well as other filled products. For further information on manufacture and operation see https://www.youtube.com/watch?t=61&v=W9iugFMDDvU.

7.2.2.4 Sheeting, Filling, and Forming

Bun production begins with sheeting of the dough. The dough sheet is then rolled into a continuous cylinder as it passes down a conveyor belt and filling is extruded into the center of the roll as it is being formed (Fig. 7.7). The cylinder of dough with filling is then passed through a cutter and former die to form the individual buns.

Improvements in modern steamed bun manufacturing equipment have incorporated novel solutions for dough forming and filling. The inclusion of laminating steps in the sheeting process has significantly improved dough handling, texture, and overall quality. For a demonstration of sheeting, filling, and forming production see http://www.yangjenq.com/equipments/equipments12.aspx?id=52.

FIGURE 7.6 Anko SD 97W automatic encrusting and forming machine. *Courtesy Anko Food Machine Co. Ltd.*

FIGURE 7.7 Machine sheeting, filling, and forming. *Courtesy Yang Jenq Machinery (Wuxi) Co. Ltd.*

7.2.3 Molding Steamed Rolls

7.2.3.1 Molding by Hand

The dough is first rolled into a thin rectangular sheet; then a layer of filling is spread over the dough sheet, which is then rolled up into a cylinder like a "Swiss roll." Individual rolls are cut from the dough cylinder. Steamed rolls can be divided into three types: rolled rolls, folded rolls, and stretched rolls (Huang, 1999).

*7.2.3.2 Rolled Rolls (*Juan Huajuan*)*

To make the scallion roll *(chongyu huajuan)* illustrated in Fig. 7.8, the thin rectangular dough piece is spread thinly with vegetable oil and sprinkled with chopped shallots (Fig. 7.8A). The dough is rolled lengthways (Fig. 7.8B) to form a long cylinder (Fig. 7.8C) that is cut crosswise into small pieces (about 35–70 g). These individual rolls may be steamed as is or molded into a number of different but special shapes by a technique of flattening and twisting the pieces so that the layers inside the rolls are exposed (Fig. 7.8D–F) (Huang, 1999). Rolled rolls are also available in many different shapes and flavors.

FIGURE 7.8 Hand making rolled rolls: (A) prepared dough with scallions, (B) rolling dough into a cylinder, (C) cutting individual pieces, (D) dough piece ready for twisting, (E) flattening dough piece with a chop stick, (F) twisted roll ready for steaming.

For example, the following references demonstrate the formation of different shaped rolled rolls: www.youtube.com/watch?v=0H1Uk4-6cKw; www.youtube.com/watch?v=N2QgXcfEORo; www.youtube.com/watch?v=5RuWAykbCzM; and https://www.youtube.com/watch?v=fYulN_wAErw.

7.2.3.3 Double Dough Rolls

These rolls are manufactured using two different doughs. Both doughs are individually rolled into similar sized thin rectangular sheets. There must be a visual difference between the two doughs which is usually achieved by using different colored and/or flavored ingredients in the second dough. The second dough is then laid on top of the first dough and the combined sheet is rolled into a cylinder from which the individual pieces are cut (see https://www.youtube.com/watch?v=GSN0C0V9saM). These layered products can also be made by machine (see Section 7.2.4).

7.2.3.4 Folded Rolls (Zhedie Juan)

These products are produced using multiple layers of dough. Lotus leaf rolls (*gua bao, gwa pao*) (Huang, 1974), popular in Taiwan and many cities in the West, are formed from two circular dough sheets placed one on top of the other. The sheets are then rolled to form a circular piece approximately 9 cm in diameter. The pieces are folded in half before steaming. They are typically served with a pork-based filling but any combination of ingredients may be used (Fig. 7.9).

Multilayer ("one thousand layer") rolls are made using a technique similar to that for making puff pastry. The dough is first rolled into a large square about 5 mm thick and then cut into half lengthways. The first piece is brushed with vegetable oil and sprinkled with flour, and the second piece is placed on top. The rectangular shape is then rolled back into a square 5 mm thick. The process is repeated a total of five times producing a square dough sheet containing 32 layers. The final dough piece is then rolled into a 4-cm-thick cylinder and cut into individual pieces (Anon., 1989; Huang, 1999). For further information, see http://www.shipinfull.com/article/26469.html.

7.2.3.5 Stretched Rolls (Chen Huajuan)

This type of roll is made from dough stretched into noodle-like strands (*la-mian*) as shown in Fig. 7.10. The strands are well oiled and then cut to the length of the roll. Dough made according to the same formula is then rolled out into a rectangle and the oiled noodles placed onto the sheet which is then wrapped into a tight package before steaming (http://www.v1.cn/2014-03-09/1025598.shtml).

One slightly different but popular example is *lup cheong guen* where a thin strip of dough is twisted around a Chinese sausage (*lup cheong*) before proofing and steaming (Fig. 7.11 and https://www.youtube.com/watch?v=XyMXL3Q8qnY).

FIGURE 7.9 Taiwanese *gua bao* (A) steamed before filling (B) resteamed and filled.

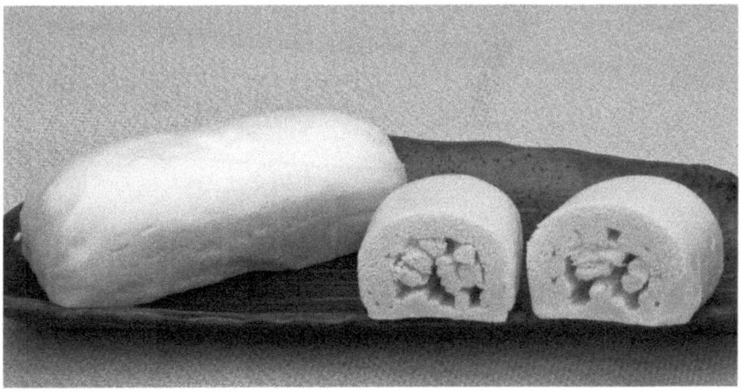

FIGURE 7.10 Steamed stretched rolls showing noodle strands inside.

7.2.4 Combined Sheeting and Molding

Small and medium-sized manufacturers may use a combination of sheeting and molding to produce quality steamed products. Doughs are first sheeted through a dough break (Fig. 7.12), before final forming and shaping. Many automated systems include a sheeting and/or laminating step before forming and molding.

FIGURE 7.11 *Lup cheong guen.*

FIGURE 7.12 Dough being sheeted through a dough break.

Two-color (double dough) rolled rolls commonly marketed as *mantou* or *roti sepan* originated in Malaysia but are now more widely available. Two doughs are mixed separately, one plain and one as in this example, with added cocoa. Each dough is sheeted separately to the same dimension and thickness. The chocolate dough is then placed on top of the plain dough and the two doughs are resheeted to the required thickness (Fig. 7.13). The dough sheet is then introduced into the molding machine which rolls the sheet longitudinally into a continuous tube (Fig. 7.14). A cutter then cuts the formed tube into individual pieces (Fig. 7.15) ready for proofing and steaming.

7.3 PROOFING

Correct proofing is crucial for the manufacture of quality steamed products. During intermediate proofing, the dough becomes more relaxed. During final proofing the dough not only relaxes, but also expands due to the carbon dioxide released during fermentation which creates the internal structure in the dough piece.

Commercial batchwise proofers are typically controlled temperature and humidity rooms or cabinets of varying size. Dough pieces are loaded onto trays and stored on racks in the proofer (Fig. 7.16). Very-large-scale facilities may install equipment to enable continuous production. Fig. 7.17 shows a tunnel proofer where residence time in the tunnel provides sufficient proofing.

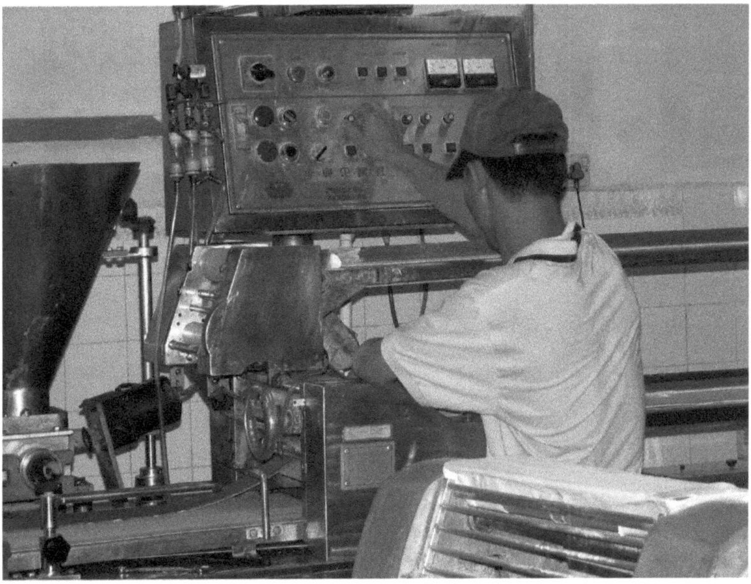

FIGURE 7.13 Resheeting two-color dough.

FIGURE 7.14 Forming two-color dough sheet into a cylinder and cutting individual pieces. *Courtesy M. Q. Lu.*

FIGURE 7.15 Two-color *mantou* after forming being loaded onto a tray for proofing.

7.3.1 Control of Proofing Conditions

Homemade steamed products are often left in steamers at ambient temperature and humidity so proofing times will vary according to ambient conditions. For industrial proofers it is essential that proofing time, temperature, and humidity are all carefully controlled. Proofers should have good internal air circulation so that they maintain even temperature and humidity throughout.

FIGURE 7.16 Proofing of steamed products in a proofing chamber. *Courtesy Dr. H. Sun and Mr. Z. Yang.*

FIGURE 7.17 Continuous proofing tunnel. *Courtesy Dr. H. Sun and Prof. K. Zhu.*

7.3.1.1 Temperature

High temperatures accelerate yeast fermentation, and may cause surface cracking if the products are overproofed. Low temperatures will prolong the proofing time, causing a flattened appearance after steaming. The optimum proofing temperature is usually 38–40°C.

7.3.1.2 Relative Humidity

Humidity can be controlled in the proofer using a fine water spray or steam injection coupled with controlled heating. The optimum relative humidity in the proofer is 70–90%. Below 60%, a skin will form and the expansion of the dough pieces is inhibited. This could cause problems with appearance and surface cracking. If the humidity is too high (over 95%), condensation will occur, resulting in a wet and sticky surface and after steaming, products may have a dull color and rough surface.

7.3.1.3 Proofing Time

Underproofed doughs will result in products with a small volume and a dense crumb texture. Products manufactured from overproofed doughs may have a flat shape, dull and rough skin, and an open and rough cell structure and are prone to shrinkage. Products manufactured by the two-step procedure need a shorter proofing time (about 60 min), while products prepared using the one-step dough procedure need a longer proofing time, in the range of 50–80 min. Some adjustments to yeast level, temperature, and/or water addition may be required to achieve a proofing time within the desired range.

7.3.1.4 Judging Correct Proofing Time

Judging the correct proofing time is a skill based on experience. The time depends on a number of factors, which are detailed in Table 7.1. Volume and appearance are the two keys to assessment of correct proofing time. For northern-style steamed bread, this is a 1.5–2.5 times increase in volume after molding. Southern-style steamed bread requires a three times increase. Adequately proofed northern-style dough piece should be expanded, and have an upright shape and a smooth and somewhat translucent surface, which should be soft and elastic to the touch.

7.4 STEAMING

7.4.1 Introduction

Proofed dough pieces can be steamed in bamboo or metal baskets over boiling water. Industrial manufacture of steamed products usually use large steaming cabinets or chambers. Steamers can be either low pressure (atmospheric) where steam is constantly vented during the cooking process or high pressure where the chamber is operated at a pressure higher than atmospheric pressure. Low-pressure steamers operate at 100°C and may require longer steaming times but they are inherently safer to operate. The operating temperature in high-pressure steamers is greater than 100°C and directly proportional to the steam pressure above atmospheric pressure. High-pressure steamers cook faster than low-pressure steamers and their operation is automatically controlled with respect to steam pressure and more importantly venting at the end of the cycle. Steaming tunnels are used for continuous production where steaming time is controlled by the speed of the product through the tunnel.

TABLE 7.1 Factors Determining Proofing Time

Factors Determining Proofing Time	Details
Type of product	Different proof times may be required for different products
Softness of dough piece	A soft dough requires a longer proofing time than a firm dough as the soft dough has higher extensibility than the firm dough. When a soft dough piece is slightly overproofed, its volume increases, but texture is still retained. If a firm dough piece is slightly overproofed, its texture becomes rough and open
Gluten content and quality	Dough made from flour with a high gluten content usually has better gluten quality and is more tolerant to overproofing than the dough made from weak flour. Dough prepared from flour with strong dough properties will need longer proofing time than flour with weak dough properties
Fermentation of dough	If the dough has undergone long fermentation, it needs a short proofing time but has less tolerance to overproofing. Dough prepared using the "no time dough" procedure needs a long proofing time and is more tolerant to overproofing
Steaming techniques	If steaming is carried out in a steaming chamber and the dough pieces are not properly proofed, quality faults such as small volume and dense texture will be evident. If the dough is steamed over a wok with cold or warm water, it takes much longer time to reach boiling temperature, and the fermentation is still occurring prior to steaming. In this case, a shorter proof time is required

7.4.2 Changes in Temperature

During low-pressure steaming, air is expelled from the steamer which facilitates the transfer of energy from the steam by thermal convection and conduction. Shortly after the beginning of the steaming cycle, the dough surface is hotter than the center of the dough piece. As steaming progresses, heat is transferred from the surface toward the center of the dough piece. Gradually the temperature in each part of dough piece reaches 100°C. The time taken for dough pieces to reach the maximum temperature in both the laboratory and commercial steamers shown in Fig. 7.18 is about 12 min, but this will depend on the size of the dough piece. To facilitate maximum steaming, the steaming chamber is vented to allow steam to escape.

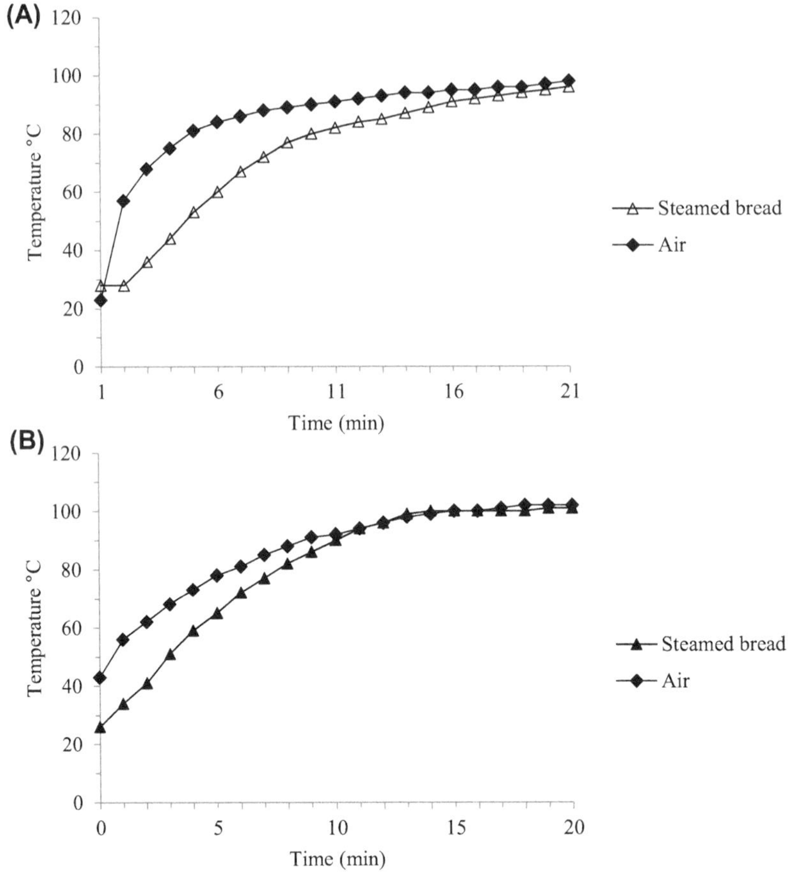

FIGURE 7.18 Temperature (°C) of air and middle of steamed bread in a laboratory (A) and commercial steamer (B).

7.4.3 Changes in Moisture

During pan bread baking, starch gelatinization commences at about 60°C as free water and protein-bound water move to the starch. The dough surface begins to lose water by evaporation almost as soon as the dough is put into the baking oven and continues throughout the remainder of baking. Most of the water loss is from the top 10mm below the crust. The total water loss is about 10% of dough weight (BRI, 1989). During steaming, there is moisture redistribution on the surface of the steamed bread. The steam actually increases the moisture content of the dough, while increasing the bread temperature accelerates evaporation of moisture from the surface.

The moisture content of the steamed bread increases, particularly during the first 10min. Initially, the moisture content is the lowest in the center and the highest on the surface. On completion of steaming, the moisture distribution becomes more even (see Section 11.5).

7.4.4 Changes in Volume

Huang (unpublished data) used a bamboo steamer and observed that the volume of steamed bread increases most quickly in the first 5 min of steaming, after which it slows. This is analogous to the "oven spring" observed in Western bread baking when a cumulative increase in gas pressure occurs due to expansion of gas cells, liberation of carbon dioxide, and vaporization of alcohol (Pyler, 1973). A similar "steamer spring" phenomenon occurs with steamed bread dough during steaming (Rubenthaler et al., 1992).

During the steaming process, gas cells expand and pressure increases on the cell walls. Carbon dioxide is liberated and migrates causing the rapid increase in volume that occurs around 40°C. As the temperature rises to around 79°C, which is achieved after about 7–9 min, there is a further increase in volume due to evaporation of alcohol. Yeast activity is also important as carbon dioxide is generated until thermal deactivation at about 60°C.

7.4.5 Formation of Skin

In the moist, hot conditions in the steamer (100°C) a thin white skin is formed on the surface of the steamed products. Skin formation occurs in the later stages of steaming, once the starch has gelatinized (Liu, 2005). This skin is very elastic and leathery. It was also observed (Huang, unpublished data) that if small bubbles appear and grow to different sizes on the surface of the steamed bread during the first 5 min of steaming, it is due to air bubbles trapped during mixing and not removed by subsequent sheeting. These small bubbles rupture when the steamer door is opened, leaving blisters on the skin of the steamed bread.

7.4.6 Control of Steaming Conditions

The steaming rate is very important to the quality of steamed bread (Chen, 2007; Huang et al., 1993). When the steaming rate is too low, the dough will not be fully cooked and the steamed bread will have a smaller volume, flat shape, poor texture, and sticky eating quality. If the steaming rate is too high, blisters, dimples, and jelly spots on the surface of the steamed bread and shrinkage can occur (see Chapter 10). These problems are more obvious using flours with high dough strength.

For industrial production, the steaming pressure needs to be optimized. Liu (2005) recommended a steaming pressure between 0.01 and 0.06 MPa. Differences in optimum steaming pressure occur because of different steaming equipment and conditions and a larger product throughput in industrial production.

Steaming times will vary according to the type of product, type of steamer, size, and amount of dough in the steaming cabinet. Filled buns, particularly those containing meat, require longer steaming time. Generally, 26–27 min steaming time was sufficient for 135 g dough pieces if the steaming pressure was between 0.02 and 0.04 MPa (Liu, 2005).

For best results, optimum steaming pressure should be reached quickly (within 2 min) once the proofed dough pieces are placed in the steamer. Insufficient steaming rate for a long period of time will actually prolong the proofing time and result in quality faults (see Chapter 10).

7.4.7 Steaming Equipment

In industrial production, steam for the steaming cabinets is generated from external boilers. Modern steaming cabinets use programmable logic controllers to regulate temperature, pressure, and time, allowing control of the batch steaming process.

7.5 COOLING AND PACKING

After steaming, the temperature of the product must be reduced before packaging. The temperature must be low enough to prevent condensation forming either on the product or in the package. The cooling rate depends on the temperature difference between the product and room temperature, relative humidity, air flow to remove hot and humid air, product size, and spacing. In most factories, products are cooled at ambient temperature without controlled cooling systems, making control of temperature and relative humidity difficult. Very low relative humidity during cooling may cause the product skin to lose moisture quickly which can induce splitting and cause the texture to become firmer.

Steamed products are packed either to protect them during freezing or to prevent contamination and damage before sale. For fresh products, packaging materials such as low-cost polyethylene bags, which may have a few small holes to prevent condensation, are a popular choice. Packaging by machine requires flexible and sealable packaging films which should possess good moisture and gas barrier properties. In frozen applications, products may first be placed in a plastic tray to both protect the product and to assist entry into the packaging machine.

7.6 CONCLUSION

New equipment is now available for most stages of steamed bread production enabling a greater throughput and production scheduling. However, correct dough handling and management of processing conditions are important in mechanized production to achieve the best final product quality.

REFERENCES

Anon., 1989. Techniques for Making Wheaten Foods, eleventh ed. Chinese Commercial Publishing House, Beijing, China (In Chinese).

BRI, 1989. Australian Breadmaking Handbook. Bread Research Institute of Australia and TAFE Educational Books, Sydney, NSW, Australia, pp. 33–65.

Chen, C.C., 2007. Taiwanese Style Steamed Bread Quality as Influenced by the Gluten Strength and Steam Generation Rate (M.Sc. thesis). University of Dayie, Taiwan (In Chinese, English abstract).

Huang, S.H., 1974. Chinese Snacks. Wei-Chuan Publishing Co., Ltd, Taipei, Taiwan.

Huang, S., Betker, S., Quail, K., Moss, R., 1993. An optimised processing procedure by response surface methodology (RSM) for northern-style Chinese steamed bread. J. Cereal Sci. 18, 89–102. http://dx.doi.org/10.1006/jcrs.1993.1037.

Huang, S., 1999. Wheat products: 2. Breads, cakes, cookies, pastries, and dumplings. In: Catharina, Y.W., Liu, K.S., Huang, Y.W. (Eds.), Asian Foods Science & Technology, first ed. Technomic Publishing Co Inc., Lancaster, PA, USA, pp. 71–108.

Huang, S., Miskelly, D., 1991. Steamed bread – a popular food in China. Food Aust. 43, 346–351.

Liu, C.H., 2005. Production Techniques of Steamed Products. Chemical Industry Publisher, Beijing, China (In Chinese).

Pyler, E.J., 1973. third ed. Baking Science & Technology, vol. 2. Siebel Publishing Company, Chicago, IL, p. 1240.

Rubenthaler, G.L., Pomeranz, Y., Huang, M.L., 1992. Steamed bread. IV. Negative steamer-spring of strong flours. Cereal Chem. 69, 334–337.

Chapter 8

Frozen Steamed Products and Frozen Doughs

8.1 INTRODUCTION

8.1.1 Frozen Steamed Products

In the past 20 years, frozen manufactured steamed bread, buns, and rolls have become readily available in supermarkets in China and East and Southeast Asia. Because of the long shelf life (often 12 months or more), frozen steamed products are also exported to markets in the United States, Europe, and Australia. These are often high-quality, premium value products. They can be stored in a home freezer and readily reheated in a steamer or microwave before being eaten for breakfast or as a snack.

8.1.2 Frozen Doughs

In the past decade, frozen dough technology, as applied to uncooked steamed products, is being utilized in China and elsewhere. This has focused on the central manufacture of preproofed and unproofed frozen doughs which are then transported to retail outlets and chain stores. Frozen doughs are not considered suitable for domestic use. Products are steamed on demand to provide fresh buns for breakfast and snacks in cities like Shanghai and Hangzhou. Use of frozen dough saves the retailers the need to purchase costly raw materials and equipment and employ highly trained staff. A space saving of 50% can be achieved in a steamed bun outlet by using frozen dough in preference to normal production (Yang, 2007).

8.2 PRODUCTION OF FROZEN DOUGH

Frozen dough for steamed bread production can be preproofed, that is, frozen and ready for steaming, or can be frozen unproofed, which necessitates thawing and proofing by the end user before steaming. The preparation for sale in a retail outlet is fast and convenient when preproofed frozen doughs are used.

Steamed Breads. http://dx.doi.org/10.1016/B978-0-08-100715-0.00008-2

Much of the information on frozen dough production has evolved from its application in Western bakery products. Frozen dough tends to have a limited storage time in the holding freezer as the yeast becomes inactivated and entrapped carbon dioxide is slowly released from the dough (Cauvain, 2014). Limited research has been undertaken to date on the applicability to steamed products and quality impact on the final products, but the smaller steamed bread portions (less than 80 g) are akin to Western bread rolls which have a better freeze–thaw tolerance than larger loaf breads.

8.2.1 Ingredients

8.2.1.1 Flour

Protein and flour strength requirements for frozen steamed bread doughs are not as high as for frozen Western breads, but the dough still needs to be able to withstand the freeze–thaw process. Freezing of dough damages the gluten network and consequently flours with high protein and high dough strength are used for frozen Western bakery products (Rosell and Gómez, 2007; Stauffer, 1993). Frozen steamed bread dough made from flour with a protein content of 9.5–11.0% was found to have the largest regions of continuous gluten network compared with lower protein flours. Higher flour protein doughs had better resistance to the demands of the freeze–thaw process, resulting in a product with specific volume, form ratio (height/width) and texture closer to the unfrozen control (Kondakci et al., 2015). Alternatively, flour protein content can be increased by the addition of dry gluten to the formulation, but better functionality at a lower cost is obtained by using flour at the correct protein content.

8.2.1.2 Yeast

Yeast activity and viability decrease when dough is frozen and a slow freezing rate is preferable. Yeast can be damaged by ice formation during freezing and the release of compounds such as enzymes which reduce gas formation (Cauvain, 2014). Cream or compressed yeast or active dried yeast (but not instant dry yeast) can be used in formulations and higher levels may be needed depending on how long the dough is to be stored in its frozen state. Fresh yeast is a cost-effective option. Yeast specifically developed for frozen dough applications has become available (http://en.angelyeast.com). As there are a number of types, addition rate and usage must be according to the manufacturers' recommendations. Some of the commercially available yeasts are very stable and have a long shelf life compared with fresh yeast provided they are stored correctly.

8.2.1.3 Other Ingredients

Changes occur in frozen doughs during storage which affect the product after steaming. Steamed bread specific volume, spread ratio, and crumb softness decrease (Kondakci et al., 2015; Sze-Yin and Lai-Hoong, 2013). Steamed breads have a more flattened profile, which is not preferred by consumers. To minimize these changes, fats and emulsifiers can be added to the formulation to retain small bubble formation in the dough, improve crumb softness, and give a finer crumb texture (see Section 4.11). Together with sugar at low levels, they act as stabilizers in frozen dough. Some functional or microingredients can also be useful. Oxidizing agents such as ascorbic acid (see Section 4.10) are commonly used in frozen baked products. They help strengthen the gluten network and aid gas retention and fermentation tolerance. A combination of trehalose (a cryoprotectant) at an addition rate of 0.1% plus maltodextrin at an addition rate of 2% was found to stabilize the quality of steamed bread made from frozen dough and the quality was equal to an unfrozen control over a 3-month storage time (Sze-Yin and Lai-Hoong, 2013). Addition of 2–4% glycerol has been found to reduce formation of ice crystals during storage (Huang et al., 2011) and improve final product quality (Ren, 2013). However, addition of the hydrocolloids xanthan gum and sodium alginate to frozen dough did not improve steamed bread quality or softness and at higher addition rates caused a decrease in specific volume and crumb softness (Kondakci et al., 2015). This is in contrast to their reported function in baked breads and even fresh steamed buns (Sim et al., 2011).

8.2.2 Frozen Dough Processing

Typical methods for production of unproofed and preproofed frozen dough are shown in Fig. 8.1. Processing conditions are very important to achieve quality salable products.

Doughs for frozen products should be mixed to optimum gluten development as described in Section 6.3. Rosell and Gómez, 2007 reported that it may be advantageous to have a lower finished dough temperature of about 20°C, and typical steamed bread dough temperatures are usually in the range of 20–24°C. Dividing and molding are carried out by one of the methods used in Chapter 7, while maintaining cool dough temperatures.

After dividing and molding the dough pieces are fermented before further processing. The fermentation conditions (see Fig. 8.1) differ depending on whether the dough piece will be preproofed or unproofed. After fermentation, preproofed dough pieces are cooled before being frozen. Unproofed dough pieces may be rested before being frozen.

Finished dough pieces are conveyed or transferred directly to the freezer. Blast freezers are generally employed to ensure the finished dough pieces are rapidly cooled with refrigerated air at temperatures no lower than −35°C to minimize yeast damage (Cauvain, 2014) and maintain good gluten structure and a high specific volume (Kondakci et al., 2015). Freezing of dough or bakery products containing 1–2% salt occurs at approximately −5°C. Because steamed bread contains virtually no added salt, the freezing point will be higher than salted doughs, closer to −1 to −2°C. As dough has poor heat conductivity the freezing process takes 30–40 min for a dough piece of less than 80 g (Liu, 2005). The rate of freezing has an impact on product quality and the amount of freezer damage that occurs.

Frozen dough pieces should be packed and sealed before transfer to the holding freezer using materials that can withstand low temperatures, protect the dough from damage, and prevent moisture loss. Frozen doughs are usually stored at −15 to −20°C. The maximum storage time has been reported to be

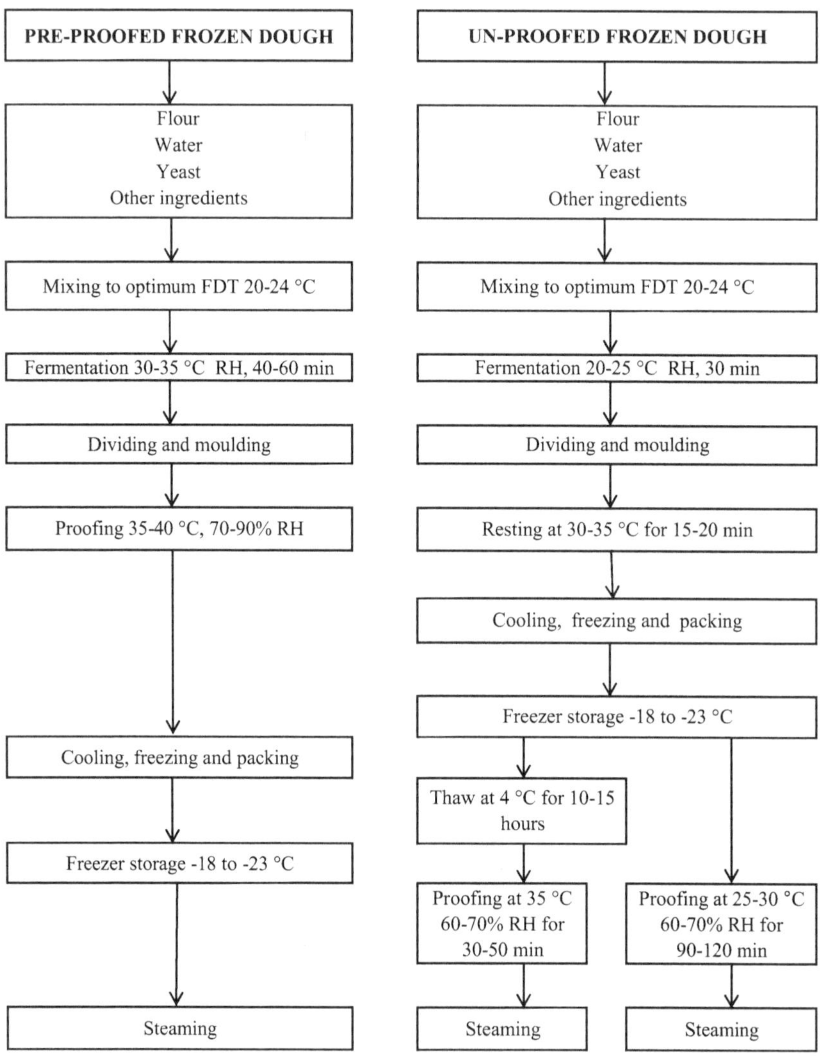

FIGURE 8.1 Production of preproofed and underproofed frozen doughs. *FDT*, finished dough temperature. *Data adapted from Liu, C.H., 2005. Production Techniques of Steamed Products. Chemical Industry Publisher, Beijing, China (In Chinese).*

3 months (Sze-Yin and Lai-Hoong, 2013), but is much shorter in commercial practice. Maintenance of the cold chain in transfer from point of manufacture to retail storage is crucial to minimize temperature fluctuations and prevent partial thawing of dough.

Preproofed doughs can be taken directly from the freezer to the steamer after which they are ready for retail sale. Steaming time is around 15–20 min for large (60–100 g) products and 10–15 min for small (30–60 g) pieces.

Unproofed frozen doughs can be either thawed at 4°C for 10–15 h and then proofed at 35°C for 30–50 min, or simply proofed at 25–30°C for 90–120 min prior to steaming (see Fig. 8.1 for the detailed procedure). The lower relative humidity at this stage helps to prevent condensation on the surface of the products.

8.3 PRODUCTION OF FROZEN FINISHED BREADS, BUNS, AND ROLLS

Frozen steamed bread, buns, and rolls are readily available for purchase and a wide variety is available. Fig. 8.2 shows some of the products available for purchase in Sydney, Australia. These products are formed, proofed, steamed, frozen, and packaged by the manufacturer before being delivered frozen to the retail trade. They are prepared for consumption by reheating by microwaving or steaming, without the need for thawing or proofing. The time taken for preparation is detailed by the manufacturer on the wrapper and will depend on the size and shape, and if the product is filled or unfilled.

The products are manufactured using normal production methods (Chapter 7). Dough lamination and sheeting are often used to ensure a high-quality product. Products are molded, filled, proofed, steamed, and then cooled prior to packaging and blast freezing. Provided the process is controlled correctly, the

FIGURE 8.2 Frozen steamed products purchased from an Asian supermarket in Sydney.

appearance, acceptability, and freshness of the frozen product on resteaming or microwaving will be similar to that of the freshly made product.

These products can be bulk packed into plastic bags or wrapped in flexible films. Many products are placed in plastic trays prior to overwrapping to add extra protection and assist the wrapping procedure. The packed product is then blast frozen. The packaging material protects the products from damage and reduces moisture loss during freezing.

During freezing of steamed products, the center of the product needs to reach about −18°C as soon as possible. However, breads stale most rapidly in the region +10 to −6°C, reaching a peak rate of staling at about 2°C (BRI, 1989). Steamed products need to be cooled through this critical range as rapidly as possible, after which a slower rate can be used. Optimal freezing is achieved using a blast freezer with good air circulation at −18°C. Freezing time is usually 1–1.5 h, depending on the freezing rate.

As with frozen dough, it is important to maintain the cold chain and minimize temperature fluctuations from production to retailer. In this case, the point of sale may be another country, and involve the use of refrigerated containers during a sea voyage in addition to land transport.

8.4 CONCLUSION

Frozen dough production and sales should increase in the future. Although current sales are mainly to convenience stores and retail chains, it is expected that smaller retailers which have a low demand for frozen dough at present will recognize and utilize the associated benefits. This will help offset labor and skills shortages as expertise will be concentrated in larger production facilities which will include lines for frozen dough production. Sales and availability of finished frozen products will continue to increase, with good potential for export sales.

REFERENCES

BRI, 1989. Australian Breadmaking Handbook. Bread Research Institute of Australia and TAFE Educational Books, Sydney, NSW, Australia, pp. 109–122.

Cauvain, S.P., 2014. Frozen dough and par-baked products. In: Zhou, W., Hui, Y.H. (Eds.), Bakery Products Science and Technology, second ed. Wiley Blackwell, Chichester, West Sussex, UK, pp. 523–537.

Huang, L., Wan, J., Huang, W., Rayas-Duarte, P., Liu, G., 2011. Effects of glycerol on water properties and steaming performance of prefermented frozen dough. J. Cereal Sci. 53, 19–24. http://dx.doi.org/10.1016/j.jcs.2010.07.006.

Kondakci, T., Zhang, J.W., Zhou, W.B., 2015. Impact of flour protein content and freezing conditions on the quality of frozen dough and corresponding steamed bread. Food Bioprocess Technol. 8, 1877–1889. http://dx.doi.org/10.1007/s11947-015-1544-6.

Liu, C.H., 2005. Production Techniques of Steamed Products. Chemical Industry Publisher, Beijing, China (In Chinese).

Ren, B., 2013. Study of Frozen Dough Production Technology (M.Sc. thesis). Xihua University, Chengdu, China, p. 74 (In Chinese, English abstract).

Rosell, C.M., Gómez, M., 2007. Frozen dough and partially baked bread: an update. Food Rev. Int. 23, 303–319.

Sim, S.Y., Noor Aziah, A.A., Cheng, L.H., 2011. Characteristics of wheat dough and Chinese steamed bread added with sodium alginates or konjac glucomannan. Food Hydrocoll. 25, 951–957. http://dx.doi.org/10.1016/j.foodhyd.2010.09.009.

Stauffer, C.E., 1993. Frozen dough production. In: Kamel, B.S., Stauffer, C.E. (Eds.), Advances in Baking Technology. Blackie Academic & Professional, Glasgow, UK, pp. 88–106.

Sze-Yin, S., Lai-Hoong, C., 2013. Effects of maltodextrin and trehalose on the physical properties of Chinese steamed bread made from frozen doughs. Int. Food Res. J. 20 (4), 1529–1535.

Yang, Z., 2007. Technology for frozen dough for steamed products and the Chinese traditional food industrialization. J. Grain Food Ind. 14 (5), 9–10 (In Chinese).

Chapter 9

Laboratory Preparation and Evaluation of Steamed Bread

9.1 INTRODUCTION

Laboratory-scale preparation of steamed bread is essential to determine flour quality requirements, evaluate changes in formulations and processes, assess new or novel ingredients, and determine the suitability of new wheat varieties and crossbreds for steamed bread production. Assessment of steamed bread quality has direct application for quality assurance in flour mill laboratories, for product development and research, and for phenotyping of quality traits in breeding programs.

Analogous to test baking of Western bread, test methods for steamed bread preparation need to be well documented. Methods must be relevant, reliable, and reproducible and be capable of discriminating between samples. Appropriate evaluation criteria must be included for the assessment of quality characteristics. Generally, a small sample size is used (100–300 g flour) and a control flour sample is included in each batch to ensure the reliability of the method and to identify unusual variability.

A review of methods published in Chinese and English has shown that many authors did not acknowledge the different styles of steamed bread, and failed to specify ingredients and/or control process conditions sufficient to enable other researchers to reproduce their test methods. Some also used Western bread formulations in place of steamed bread formulations or used excessively high levels of water addition based on Western bread test baking. Many did not appreciate the wide variety of products in the market or the range of geographic preferences.

9.2 LABORATORY PROCESSING METHODS

There are numerous laboratory methods documented by a diverse range of researchers for the assessment of different styles of steamed breads. Some methods are based on Western bread-making principles, while others are established more on traditional procedures. Often the method was devised using equipment

Steamed Breads. http://dx.doi.org/10.1016/B978-0-08-100715-0.00009-4

already existing and readily available in the laboratory. There seems to be no general agreement as to processing conditions. It is often unclear whether the methods have been optimized or validated against other methods. The most popular "base" methods are the Chinese Standard Methods SB/T 10139-93 (1993) and GB/T 17320-1998 (1998) and Huang et al. (1993, 1998). A number of researchers have modified these methods by changing either the formulation and/or the processing conditions.

9.2.1 Chinese National Standard Procedures

Chinese standard SB/T 10139-93 specified making steamed bread entirely by hand. A new method, GB/T 1730-1998, was issued in 1998. Although hand kneading was mostly retained in the new method, the initial processing of the dough specifies mixing the ingredients for 1 min in a Swanson pin mixer. The 1993 method is a straight dough procedure with relatively short proofing time. In contrast, the 1998 method is a two-step (essentially a modified sponge and dough) method with an extended proofing time (Table 9.1). A modification of the 1993 standard method using a mechanical mixer and controlled fermentation and proofing conditions (Sha et al., 2007) is also shown for comparison.

Considering the ingredients listed for the methods detailed in Table 9.1 and the associated scoring systems outlined in Table 9.4, the authors conclude that these methods relate to the assessment of flour suitable for northern-style steamed breads.

A third standard Chinese method GB/T 17320-2013 (2013) has been released to replace the 1998 method. This is essentially a one-step procedure and will be discussed in Section 9.2.3.

9.2.2 Optimized Two-Step Methods

The two-step procedure was described in Section 6.5. A portion of the flour is mixed with all the water and other ingredients and mixed before the first fermentation phase (sponge or preferment). After the preferment stage, the remaining flour is added and after remixing, the dough is sheeted before dividing, rounding, final proofing, and steaming. Early versions of the two-step method were published by Faridi and Rubenthaler (1983), Lin (1983), and Lin et al. (1990). Response surface methodology was used to develop the methods of Huang et al. (1993, 1998) and Su (2005) and these are shown in Table 9.2. Processing variables such as water addition, mixing time, fermentation time, and temperature and steaming time were optimized. Water addition is based on a percentage of farinograph water absorption, bringing it into the range normally used for steamed bread (see Section 6.3.2). Allowance was made for flours of differing mixing requirements, based on farinograph development time (Huang et al., 1993, 1998). In Su's method, dusting flour was used during sheeting.

TABLE 9.1 Comparison of Chinese National Standard Methods for Steamed Bread

	Chinese Standard SB/T 10139-93 (1993)	Sha et al. (2007) (Modified Chinese Standard SB/T 10139-93 (1993))	Chinese Standard GB/T 17320-1998 (1998)
Style	Northern	Northern	Northern
Ingredients			
Flour (g)	100	2000	90
Water (%)	48	48	75% FWA
Dry yeast (%)	1	1	1
Mixing			
Mixer	Combine to dough with glass rod or chopstick, then knead by hand	Unspecified	Swanson pin mixer for 1 min, then hand knead into a ball
Time (min)	3	4 min at 60 rpm	Kneading time unspecified
Fermentation			
Temperature (°C)	38	38	29–30
Relative humidity (%)	Unspecified	85	80–85
Time (min)	60	60	150
Hand kneading (min)	3		Fermented dough is hand kneaded 15 times with 10 g of the same flour and 0.2 g of sodium bicarbonate
Dividing and Molding			
Dough weight (g)	Unscaled	100	Unscaled
Molding	By hand	Rounded by hand	By hand
Proofing			
Temperature (°C)	Ambient	38	29–30
Relative humidity (%)	Ambient	85	80–85
Time (min)	15	25	15
Steaming			
Time (min)	20 (from when steam is first produced)	20	20

FWA, Farinograph water absorption.

TABLE 9.2 Optimized Processing Procedures for Two-Step Method for Steamed Bread From Huang and Su's Laboratories

	Huang et al. (1993)	Su (2005)	Su (2005)	Huang et al. (1998)	Su (2005)
Style	**Northern**	**Very Firm**	**Firm**	**Southern**	**Soft**
Ingredients					
Flour (g)	(200+100)[a]	(265+110+125)[b]	(350+100+50)[b]	(240+60)[a]	(450+35+15)[b]
Water	70% FWA	65% FWA	72% FWA	80% FWA	80% FWA
Dry yeast (%)		1	0.8		0.5
Fresh yeast (%)	1.5			1.5	
Mixing					
Mixer	Farinograph	TP101[c]	TP101[c]	Farinograph	TP101[c]
Mix time (min)	75% FDDT	4	4	50% FDDT	4
Mix speed (rpm)	60	2 min at 60 plus 2 min at 90	2 min at 60 plus 2 min at 90	60	2 min at 60 plus 2 min at 90
Fermentation (first)[d]					
Temperature (°C)	32	32	32	32	32
Relative humidity (%)	85	85	85	85	85
Time (min)	60	40	40	150	40
Remix (min)	25% FDDT	2	2	180% FDDT	2
Fermentation (second)					
Temperature (°C)		32	32		
Relative humidity (%)		85	85		
Time (min)		20	20		

Sheeting					
No. passes	20	40	20	20	20
Gap (mm)	7.2	7	7	7.2	7
Dividing and Rounding					
Dough weight (g)	100	Approx. 149	Approx. 160	100	Approx. 150
Molding	Extensograph rounder			Extensograph rounder	
Proofing					
Temperature (°C)	32	32	32	32	32
Relative humidity (%)	85	85	85	85	85
Time (min)	20	25	20	35	20
Steaming					
Time (min)	20	20	20	20	20
Time to assessment (min)	30	30	30	30	30

FWA, Farinograph water absorption; *FDDT*, farinograph dough development time.
[a]*Sponge + remix.*
[b]*Preferment + remix + dusting flour during sheeting.*
[c]*Spiral mixer.*
[d]*In Su's methods, prefermentation was used as the first fermentation.*

9.2.3 One-Step Methods

Many commercial products are manufactured using the one-step process (Section 6.5). This is the simplest and fastest method, requiring little or no fermentation after mixing and can be used in the laboratory to produce steamed bread with the additional advantage of a higher sample throughput. Sheeting is specified by two authors and this helps develop the dough and improve the structure. Although the methods detailed in Table 9.3 are for the laboratory preparation of Guangdong-style and northern-style steamed breads, the methods can be adapted for use for southern style.

9.2.4 Sourdough Procedure

In laboratory sourdough procedures, the sourdough starter replaces a portion of the flour. Wu et al. (2012a) used a formula of 100% flour, 48% water, 0.8% active dry yeast, and 30% sourdough starter mixed at low speed for 5 min and medium speed for 2 min in a Hengyue B10B mixer. Doughs were divided, rounded, molded, proofed for 30 min at 38°C and 80% relative humidity (RH), and finally steamed for 12 min. A similar sourdough addition rate of 30% was used by Keeratipibul et al. (2013), although salt and sugar were added and water addition was determined from the farinograph water absorption.

Sourdough methods can be useful in product development, or in the investigation of sourdoughs, but are probably too time consuming where a rapid throughput is required, such as in a wheat breeding program.

9.2.4.1 Char Siew Bao

Char siew bao is a special segment of Guangdong-style steamed bun and was described in Chapter 1. Preparation of the sourdough starter, which is allowed to ferment for 2–3 days, is detailed by Limley et al. (2013). The remaining process is basically a two-step or sponge and dough procedure in which mixing time, fermentation time, neutralization, and proofing and steaming are carried out under controlled conditions. The test method is applicable only to this specific product.

9.3 GUIDELINES FOR LABORATORY PROCESSING

Test method choice will depend on the intended application and purpose. Different methods may rank and discriminate samples differently and this is important for evaluation of wheat varieties and crossbred lines in a breeding program. Whichever method is used in the laboratory for small-scale processing, it will need to be optimized according to available equipment, local conditions, and product preferences. The following guidelines are designed to assist in this process.

9.3.1 Water Addition

Some methods specify fixed water addition, but optimum water addition is preferred to ensure dough of the correct consistency. This can be estimated from

TABLE 9.3 Comparison of One-Step Methods

	Kim et al. (2001)	Huang and Quail (1997)	Chinese Standard GB/T 17320-2013 (2013)
Style	**Guangdong**	**Guangdong**	**Northern**
Ingredients			
Flour (g)	100	200	200
Water (%)	Variable, depends on flour	75% FWA	75% FWA
Dry yeast (%)	1.5		1
Fresh yeast (%)		1.5	
Sugar (%)	8	8	
Shortening (%)	3	4	
Salt (%)	1		
Baking powder (%)	1	0.5	
Yeast food (%)	0.8		
Mixing			
Mixer type	Hobart N50	Pin	Unspecified[a]
Time (min)	Variable, depends on flour	To peak resistance	6
Final dough temperature (°C)	26	30	Unspecified
Resting time (min)		15	No
Sheeting			
No. passes		20	15–20
Gap (mm)		6	6
Dividing and Molding			
Dough weight (g)	40	100	Approx. 140
Molding	By hand	Extensograph rounder	By hand
Proofing			
Temperature (°C)	32	32	37
Relative humidity (%)	75	85	75–80
Time (min)	40	40	40
Steaming			
Time (min)	12	20	15

FWA, Farinograph water absorption.
[a]*Method specifies a variable speed mixer (0–200 rpm) with a bowl capacity (100–500 g).*

TABLE 9.4 Comparison of Quality Characteristics and Score Allocation in Evaluation Systems of Steamed Bread in the Literature

Method	A	B	C	D	E	F	G	H	I	J
Volume	v									
Weight	v									
Specific volume	v	10	20	15	25	25	25	20	25	15
Spread Ratio										
Height/width					5	10	10		10	
Height				5						5
Appearance and Texture										
Shape		10	7.5	10				7.5		10
Skin smoothness	10	15	7.5	10	10	10	10	7.5	10	10
Skin whiteness	5	10	10	10	5	5	5	5	5	10
Skin thickness		5								
Skin shininess	5				5	5	5		5	
Crumb whiteness	10	10			5	5	5	5	5	
Softness or hardness	v	25			0	10	5			

Elasticity and cohesiveness			20							
Cohesiveness				10	10		5	10	10	10
Elasticity or springiness				10	100	10	10	10	10	10
Structure	10		15	15	15	10	10	15	10	15
Stickiness			15	10	10	10	10	15	10	10
Taste										
Flavor			5	5				5		5
Eating quality		15								
Total score		100	100	100	100	100	100	100	100	100

v, Visual or physical quality character used, but no score was allocated.
A: Faridi and Rubenthaler (1983), Rubenthaler et al. (1990); **B**: Lin et al. (1990); **C**: Chinese standard SB/T 10139-93 (1993); **D**: Wang et al. (1998); **E**: Huang et al. (1993) (northern style); **F**: Huang et al. (1998) (southern style); and **G**: Su (2005) (soft type); **H**: Liu (2005); **I**: Liu et al. (2006) (southern style); **J**: Chinese standard GB/T 17320-1998 (1998), Chinese standard GB/T 17320-2013 (2013).

the farinograph water absorption (FWA). As the level of water addition increases from 70% to 90% FWA, steamed bread volume and spread ratio increase (He et al., 2007) and the crumb becomes more open. If the water addition is too low, the dough will be too dry and molding problems will be experienced (Huang et al., 1993). Optimum water addition is about 70% FWA for northern-style and 80% FWA for southern-style steamed bread (see Section 6.3.2).

9.3.2 Mixing Conditions

Mixing conditions will depend on available equipment in the laboratory. A farinograph mixer at 60 rpm (Huang et al., 1993), spiral mixer (Su, 2005), and a pin mixer (Rubenthaler et al., 1990) are all suitable. A variable mix time estimated according to farinograph development time (Huang et al., 1993) is preferable, but an experienced operator can also estimate optimum mixing time. Mixing time and speed will need to be optimized as part of the initial setup process according to the mixer used. Doughs should be mixed to optimum development, although there is some latitude as further dough development will occur during the sheeting phase if one is included. Author (Huang) judged the optimum mixing time by observing the dough when a pin mixer was used. When the dough begins to adhere to the bottom of mixer bowl, optimum mixing has been achieved. High-speed mixers should not be used because of the difficulty in controlling mixing time and dough temperature.

Finished dough temperature should be controlled within the range 30 ± 1°C.

9.3.3 Proofing Conditions

All fermentation and proofing should be carried out at standardized temperature (up to 40°C) and humidity (80–85% RH) using a proofing cabinet or incubator capable of maintaining ±1°C. The proofing time will vary according to the proofer conditions and experience. Objective methods to determine correct proofing time by measuring the increase in dough volume in a calibrated tube or cylinder are described by Hou and Popper (2007) and Wu et al. (2012b).

9.3.4 Sheeting Conditions

Dough sheeting will assist with dough development and give a smoother surface and finer crumb structure. For laboratory use, a small commercial dough break can be used, or preferably a noodle machine with sheeting rolls 120–150 mm diameter, 150–220 mm width, roll speed set at 9 rpm, and roll gap 6–7 mm. If the roll diameter is too small, the dough will not sheet properly. The optimum number of sheeting passes depends on the mixing time and in the two-step method, proof time as well (Huang et al., 2003). All noodle machine settings need to be calibrated and optimized. The dough sheets are folded end to end and resheeted in the longitudinal direction.

9.3.5 Dividing and Rounding

After the doughs are divided to a set weight, the common methods of rounding are either by hand or by using the rounder on the extensograph. A small commercial bun rounder may also be suitable, but will require a larger sample size.

It is important that the dough piece is only kneaded lightly during molding as differences in molding can result in variability in subsequent texture analysis tests.

9.3.6 Steaming Conditions

A commercial electric steaming cabinet with adjustable steam pressure and transparent inner door is preferable to using a bamboo steamer over a wok with boiling water as there is better control of steaming conditions. Within a laboratory, optimum steaming pressure and steaming time will need to be established depending on the size and number of dough pieces, but it usually varies between 12 and 20 min.

A steaming rate producing 14 g of condensate per minute from a 90-L steamer was used for laboratory-scale production by Huang et al. (1993). The use of a steam pressure of 0.003 MPa for 8 or more evenly sized pieces steamed for 20 min in the laboratory appears to be optimum (Huang, unpublished data). Quality deteriorated above this rate. Chen (2007) confirmed that a high steam generation rate caused shrinkage in steamed bread made from medium and high dough strength flours.

9.3.7 Time to Evaluation Tests

A variety of times before evaluation have been used by different authors. These range from 15 min to 24 h. There is a need for standardization as steamed bread stales quickly. A time lapse of about 30–60 min after steaming should be adequate to bring samples to about 22°C, at which time sensory and instrumental evaluation should be carried out. Samples should be stored at specified temperature and RH.

9.4 QUALITY EVALUATION SYSTEMS

Quality evaluation of steamed bread should be based on both objective and subjective quality assessments where each quality character must be identified and the range clearly defined such that a total score can be assigned.

A number of authors have attempted to establish quality evaluation systems for steamed bread (Faridi and Rubenthaler, 1983; Guo et al., 2002; Huang et al., 1993, 1998; Kruger et al., 1992; Lin et al., 1990; Liu, 2005; Liu et al., 2006; Chinese Standard Methods SB/T 10139-93, GB/T 17320-1998, GB/T 17320-2013; Rubenthaler et al., 1990; Su, 2005; Wang et al., 1998). Table 9.4 compares the characteristics and scores between these different methods.

Assessments can be categorized into external characters, appearance, texture, and flavor. No system includes all the characteristics listed in the table indicating that different researchers may have placed different importance to each character to develop a system suitable for their own specific research needs. Therefore it may not be possible to compare total scores using different systems unless the ranking between systems is consistent. The most common evaluation systems used in the literature are those of (C and J) Chinese standards SB/T 10139-93 and GB/T 17320-1998 and (E) Huang for northern style; and (F) Huang for southern style.

9.4.1 Characteristics

Objective characters including volume, specific volume, and spread ratio as detailed in Section 9.4.2 are the easiest to rate as they are directly measurable. Subjective characteristics of appearance, flavor, and texture are more difficult to determine as they usually require an "opinion" of an evaluation panelist. Panel training is an important aspect of assessment so as to standardize the ratings given to each character between panelists. Instrumental methods are being developed to measure some subjective characteristics. As with sensory panel training, correlation must be established between sensory score and instrumental score. Instrumental methods offer a more objective perspective and are being incorporated into quality evaluation methods.

9.4.2 Measurable Characteristics

Specific volume (volume/weight) is considered one of the most important parameters with one of the highest score allocations. Traditionally the volume was determined by rapeseed displacement. Shape is defined by the spread ratio (width/height) (see Section 2.3.2). A high spread ratio is undesirable, as it indicates a flat shape. Measurement of the spread ratio is valid only for steamed bread with a spherical shape. In the case of steamed bread with a cylindrical or log shape, height, width, and length should be measured.

9.4.3 Appearance

Appearance characteristics such as smoothness, whiteness, and shininess of the skin, and crumb whiteness are considered as important in almost all systems. These are evaluated subjectively by trained sensory panels. High scores are given to very smooth, shiny and white skin; and white crumb.

9.4.4 Texture

Crumb structure is very important and it is emphasized by all authors. Northern-style steamed bread should have a fine and even cell distribution. Steamed bread softness, cohesiveness, and elasticity are also important eating quality items. If bread sticks to the teeth when chewed, it is disliked by most consumers. Steamed bread becomes sticky when rain damaged wheat has been used, flour has high starch damage, the steaming rate is too low, or steaming time is too short. Therefore stickiness is used to guide evaluation, but not ignoring these quality faults.

9.4.5 Flavor

Flavor is considered important in all the Chinese standard methods, with a wheaty and pleasant taste preferred.

9.4.6 Scoring Systems

9.4.6.1 Northern Style

The evaluation systems for northern-style steamed bread for Chinese standard SB/T 10139-93, Chinese standard GB/T 17320-1998, and Huang are detailed in Tables 9.5–9.7. The latest Chinese standard GB/T 17320-2013 retains the 1998 scoring system.

TABLE 9.5 Quality Evaluation System for Steamed Bread Using Chinese Standard SB/T 10139-93

Quality Parameter	Maximum Score	Criterion
External Attributes		
Specific volume, sv (ml/g)	20	Score=20 for sv>2.3. When sv<2.3 score=25–(2.3–sv)*10
Appearance	15	Smooth, symmetrical, well rounded, 12.1–15; medium, 9.1–12; coarse, asymmetrical, hard spots, 1–9
Internal Attributes		
Color	10	White, creamy white, 8.1–10; medium, 6.1–8; gray, dark, 1–6
Structure	15	Fine and even, 12.1–15; medium, 9.1–12; coarse and uneven, 1–9
Elasticity and cohesiveness	20	Elastic and cohesive, 16.1–20; medium, 12.1–16; inelastic and uncohesive, 1–12
Stickiness	15	Does not stick to teeth when chewed, 12.1–15; medium stickiness, 12.1–16; sticks to teeth, 1–9
Flavor and aroma	5	Normal smell, 4.1–5; medium smell, 3.1–4; abnormal smell, 1–3
Total score	100	

Data from Chinese standard method SB/T 10139-93, 1993. Professional Standards for Specific End Uses of Wheat Flour for Steamed Bread, SB/T 10139-93. Ministry of Commerce, The People's Republic of China (In Chinese), and Zhu, F., 2014. Influence of ingredients and chemical components on the quality of Chinese steamed bread. Food Chem. 163, 154–162. http://dx.doi.org/10.1016/j.foodchem.2014.04.067 0308.

TABLE 9.6 Quality Evaluation System for Steamed Bread Using Chinese Standards GB/T 17320-1998 and GB/T 17320-2013

Quality Parameter	Maximum Score	Criterion
External Attributes		
Specific volume, sv (mL/g)	15	sv≥2.8 full score, 15; sv≤1.5 the lowest score 2; 2.8>sv>1.5; deduct 1 score for every decrease of 0.1 sv
Height (cm)	5	Height≥7.0 cm full score, 5; height≤5.0 cm the lowest score of 1; height less than 7 cm deduct 1 score for every 0.5 cm below 7.0 cm in height
Skin color	10	White, creamy color; 8–10 score; light yellow and yellow, 6–8 score; gray 2–6 score
Skin smoothness	10	Smooth, 8–10 score; shrinkage, collapsing, blisters, dimples, and jelly spots, 3–8 score
Shape	10	Even, upright, 7–10 score; flat and uneven, 4–7 score
Internal Attributes		
Structure	15	Crumb holes are small and even, 12–15 score; crumb holes are too small but even, 8–12 score; big holes and structure is rough, 5–11 score; skin separating from crumb, 8–12
Springiness	10	Springs back quickly when compressed to 50%, 7–10 score; weak or no spring back, 3–7 score; difficult to press, firm, 2–6 score
Cohesiveness	10	Strong cohesiveness, 7–10 score; weak cohesiveness, very firm, 4–7 score
Stickiness	10	Clean, does not stick to teeth, 8–10 score; sticks to teeth, 3–7 score
Flavor	5	Wheaten flavor, no off smell and taste, 4–5 score; wheaten flavor is not strong, 3–4 score; off flavor and taste, 1–3 score
Total	100	

TABLE 9.7 Quality Evaluation System for Northern-Style Steamed Bread (Huang)

Quality Parameter	Maximum Score	Criterion
Specific volume, sv (mL/g)	25	Score=25−(3.0−sv)/0.08
Spread ratio (width/height)	5	1.35–1.39 scores 5
		1.40–1.44 scores 4
		1.45–1.49 scores 3
		1.50–1.54 scores 2
		1.55–1.59 scores 1
		1.60–1.64 scores 0
External Appearance		
Skin whiteness	5	High score for very white skin (subjective and objective methods available)
Skin brightness	5	High score for very shiny skin (subjective)
Skin smoothness	10	High score for very smooth skin, free of wrinkles, dimples, blisters, or gelatinized spots (subjective)
Internal Appearance		
Crumb whiteness	5	High score for very white crumb (subjective and objective methods available)
Crumb structure	15	High score for evenly open crumb (subjective)
Elasticity and cohesiveness	20	High score for elastic and cohesive crumb (subjective and objective methods available)
Stickiness	10	High score for crumb that does not stick to teeth when chewing and with normal taste (subjective and objective methods available)
Total score	100	

Based on Huang, S., Betker, S., Quail, K., Moss, R., 1993. An optimised processing procedure by response surface methodology (RSM) for northern-style Chinese steamed bread. J. Cereal Sci. 18, 89–102. http://dx.doi.org/10.1006/jcrs.1993.1037; Huang, S., Quail, K., Moss, R., 1998. The optimisation of a laboratory processing procedure for southern style Chinese steamed bread. Int. J. Food Sci. Tech. 33, 345–357. http://dx.doi.org/10.1046/j.1365-2621.1998.00166.

TABLE 9.8 Quality Evaluation System for Southern-Style Steamed Bread

Quality Parameter	Maximum Score	Criterion
Specific volume, sv (mL/g)	25	Score = 25–(3.2–sv)/0.08
Spread ratio, (width/height)	10	1.40–1.44 scores 10
		1.45–1.49 scores 9
		1.50–1.54 scores 8
		1.55–1.59 scores 7
		1.60–1.64 scores 6
		1.65–1.69 scores 5
		1.70–1.74 scores 4
		1.75–1.79 scores 3
		1.80–1.84 scores 2
		1.85–1.89 scores 1
		>1.9 scores 0
External Appearance		
Skin whiteness	5	High score for very white skin (objective)
Skin brightness	5	High score for very shiny skin (subjective)
Skin smoothness	10	High score for very smooth skin, free of wrinkles, dimples, blisters, or gelatinized spots (subjective)
Internal Appearance		
Crumb whiteness	5	High score for very white skin (objective)
Crumb structure	10	High score for evenly open crumb (subjective)
Softness and cohesiveness	10	High score for soft and cohesive crumb (objective)
Elasticity	10	High score for elastic crumb (objective)
Stickiness	10	High score for crumb that does not stick to teeth when chewing and with normal taste (objective)
Total score	100	

Reproduced with permission from Huang, S., Quail, K., Moss, R., 1998. The optimisation of a laboratory processing procedure for southern style Chinese steamed bread. Int. J. Food Sci. Tech. 33, 345–357. http://dx.doi.org/10.1046/j.1365-2621.1998.00166.

Using the scoring system in Table 9.6, a score less than 70 is considered poor quality, a score of between 70 and 79 is normal, a score of 80–89 is good, and a score in excess of 90 is excellent.

The criteria and score weightings for Huang's northern-style system differ from the Chinese standard GB/T 17320-1998 in that there is a higher score for specific volume and spread ratio. In a comparative study of the two evaluation methods, the total scores were positively correlated ($r=0.66^{**}$) and Huang's method was better able to distinguish differences between flours with different gluten strength (Chen et al., 2010).

Huang's quality evaluation system for northern-style steamed bread was established based on consumer preference testing and investigation of quality characteristics. Similar rankings were allocated by Su (2005), who used 21 experts to investigate sensory characteristics and their relative importance to quality and analyzed the results using "fuzzy clustering" mathematics between two data sets.

In Huang's method, the total score was considered to be the best overall judgment of steamed bread quality and descriptive categories were allocated as follows: outstanding (>90), excellent (85–90), very good (80–85), good (75–80), fair (70–75), unsatisfactory (65–70), poor (60–65), and unacceptable (<60).

9.4.6.2 Southern Style

Southern-style steamed bread has a more open crumb structure than northern style and by comparison, the texture is soft and slightly cohesive. The main difference between Huang's northern and southern systems (Tables 9.7 and 9.8) is the difference in allocated specific volume scores and crumb structure, softness, and cohesiveness. These textural attributes were identified as important in consumer preference tests (Huang et al., 1998).

9.4.6.3 Guangdong Style

Guangdong-style steamed bread contains sugar or fat and sugar and the product has an open crumb structure and a soft texture which is not cohesive. There are many different types of products and many regional preferences. Filled buns and specialty buns are also common. Much of the expertise in this area resides in commercial practice and there are few references to scoring systems in the literature. However, consumer preferences would indicate that a scoring and evaluation system close to that for southern style (Table 9.8) is applicable.

9.4.6.4 Char Siew Bao

Unfilled *char siew bao* is used for quality evaluation. It should be noted that because crumb texture and appearance are quite different from all other styles, Limley et al. (2013) used a scoring system based on consumer sensory research and consultation with a local flour mill. This method differs from the method of Huang et al., 1998 (Table 9.8) in that slightly more emphasis is given to specific volume and the spread ratio was not measured. All assessments were based on objective measurements. Both methods place similar importance on external appearance and crumb texture.

9.5 OBJECTIVE ASSESSMENT OF STEAMED BREAD QUALITY

Inclusion of objective measurements of steamed bread quality is advantageous when developing and maintaining a robust quality system. Objective measurements should be included when they are shown to be more accurate and reproducible. Expert panels require training and experience, and cannot normally evaluate large sample sets. A large amount of data can be generated quickly using instrumentation, but simply applying software and techniques developed by the instrument manufacturers may not always produce data that correlate with those parameters perceived to be of importance by expert panels or consumers. Besides the physical measurements discussed earlier in this chapter, color measurements using tristimulus color meters and texture analyzers (Fig. 9.1) are now widely used in the food industry and image analysis and laser technology are becoming available.

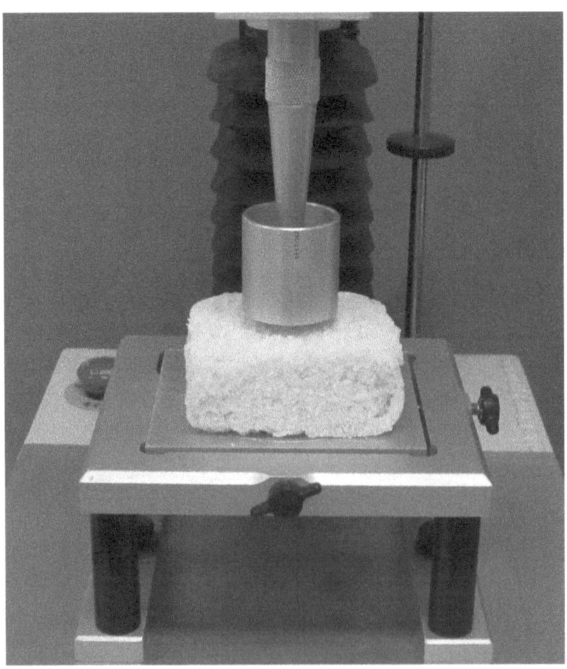

FIGURE 9.1 Texture compression test on steamed bread.

9.5.1 Texture Analysis

Deformation was the earliest method used to evaluate the texture of leavened products. Instruments measured the maximum force to compress the product a set distance, or alternatively the distance (or percentage compression) when subject to a set force. In sensory evaluation, this is often referred to as "firmness" or "softness" (Bourne, 1993). Early instruments were relatively simple and steamed bread crumb firmness was measured as maximum force using a probe attached to a rheometer (Rubenthaler et al., 1990) or an Instron Universal Testing Machine (Lin et al., 1990; Kruger et al., 1992). The development of dedicated food instruments such as SMS TA.XT (www.stablemicrosystems.com) equipped with computer software enables numerous possibilities for instrument setup, testing profiles, testing probes, and data storage and analysis.

Instrumental textural analysis must be correlated against sensory analysis before it can be used as a predictor. Huang et al. (1995) compared sensory results for northern-style steamed bread with those measured using an Instron Universal Testing Machine.

Panelists evaluated 20-mm-thick horizontal pieces obtained from the center of the bread for the following:

1. Crumb firmness—by compressing the sample between thumb and index fingers.
2. Crumb stickiness—by biting down with the molar teeth twice and assessing the force to pull the teeth apart.
3. Elasticity (springiness)—by assessing the degree of recovery on a piece fully compressed between the thumb and index finger. High scores were allocated for crumb recovery or spring back which was slow but complete with retention of the crumb texture.
4. Cohesiveness—by chewing six times with the molars and evaluating the degree to which the sample stayed together as a mass.

For the Instron test, a 28.5-mm horizontal slice was compressed 50% at a crosshead speed of 100 mm/min. Two testing profiles were used. The first was a two-cycle test or texture profile analysis (TPA). This is essentially a "two-bite" test that simulates chewing action. A typical TPA force/time curve for steamed bread is shown in Fig. 9.2. The measurements taken are described in the figure legend.

The second testing profile was a stress relaxation (SR) test (Fig. 9.3). The initial compression test was stopped for a 4-s dwell time at 50% compression (ie, at peak) and the force measured (P3). SR was defined as

$$SR = 100 * (P1 - P3)/P1.$$

Significant correlations were found between the instrumental testing and the results of the sensory panel. The maximum compression force (P1) showed significant inverse correlations with softness and cohesiveness. SR showed

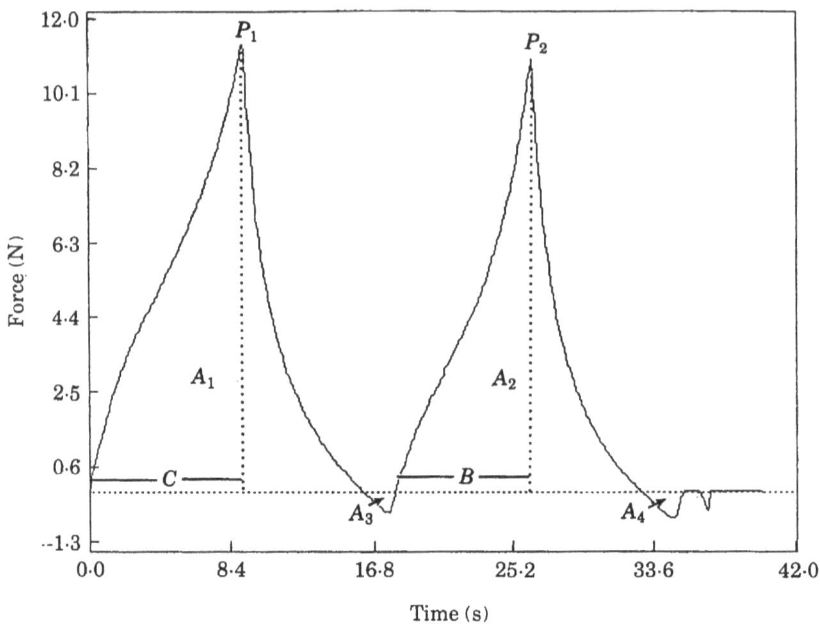

FIGURE 9.2 Instron compression test (two cycles) force–time curve. P1, peak force during the first compression cycle; P2, peak force during the second compression cycle; A1, positive force area during the first compression cycle; A2, positive force area during the second compression cycle; A3, negative force area during the first compression; A4, negative force area during the second compression; B:C, the proportion of height recovery of steamed bread during the time between the end of the first cycle and the start of the second cycle. *Reproduced with permission from Huang, S., Quail, K., Moss, R., Best, J., 1995. Objective methods for the quality assessment for northern style Chinese steamed bread. J. Cereal Sci. 21, 49–55. http://dx.doi.org/10.1016/S0733-5210(95)80007-7.*

significant inverse correlations with overall eating quality. Both the ratios A2/A1 and A3/A1 were correlated with elasticity as positive and negative correlations, respectively (Table 9.9).

From this study, it was concluded that instrumental measurement of SR was shown to be capable of replacing sensory panel softness, elasticity, nonstickiness, and cohesiveness evaluations for northern-style steamed bread. SR has been used in a number of subsequent studies for characterization of flour quality requirements (Huang et al., 1995), protein quality (Zhang et al., 2008), puroindoline alleles (Chen, 2007), and molecular markers (Xia et al., 2013).

Texture analysis is also applicable to southern-style steamed bread, using the TPA cycle shown in Fig. 9.2. The significant TPA parameters were found to be different for southern-style steamed bread, probably due to the different textural properties compared with northern style. Huang et al. (1998) allocated points to softness and cohesiveness, elasticity, and nonstickiness based on the peak force (P1) and ratios A2/A1 and A3/A1, resulting in the scoring system shown in Table 9.10.

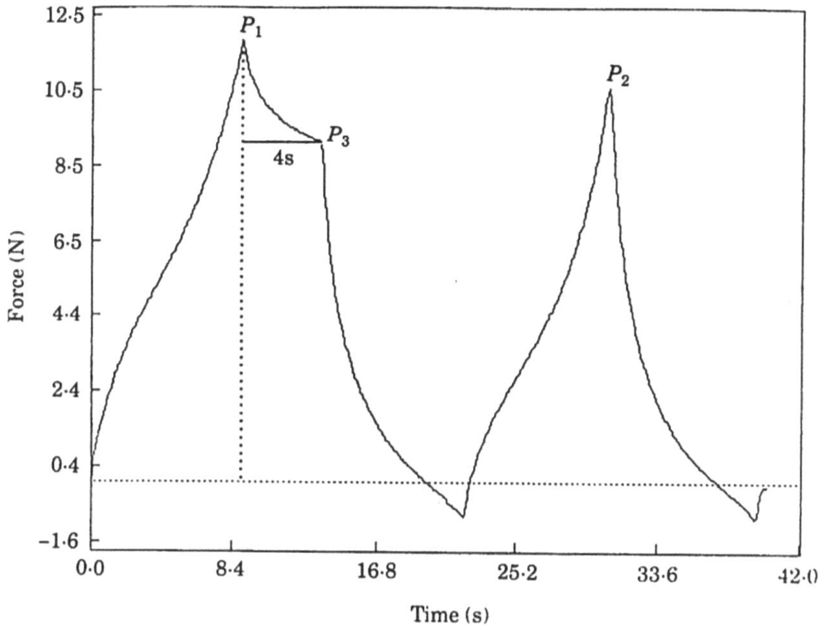

FIGURE 9.3 Instron compression test (stress relaxation) force–time curve. P1, peak force during the first cycle compression; P3, peak force after 4-s dwell time during the first compression; stress relaxation (SR) = (P1−P3)/P1. *Reproduced with permission from Huang, S., Quail, K., Moss, R., Best, J., 1995. Objective methods for the quality assessment for northern style Chinese steamed bread. J. Cereal Sci. 21, 49–55. http://dx.doi.org/10.1016/S0733-5210(95)80007-7.*

TABLE 9.9 Correlation Analysis of the Relationship Between Sensory Data and Instron Measurements for the Textural Properties of Northern-Style Steamed Bread

	Maximum Force P1	Area A2/ Area A1	Area A3/ Area A1	Stress Relaxation (SR)
Softness	−0.94**	ns	ns	ns
Elasticity	ns	0.84*	−0.82*	−0.82*
Nonstickiness	0.87*	ns	−0.77*	ns
Cohesiveness	−0.92**	ns	ns	ns
E × C[a]	ns	0.84*	ns	−0.93**
E × C × S × N[b]	ns	0.87*	ns	−0.93**

See Fig. 9.2 for sample texture profile analysis graph.
*$p<.05$; **$p<.01$. SR is defined by Huang, S., Quail, K., Moss, R., Best, J., 1995. Objective methods for the quality assessment for northern style Chinese steamed bread. J. Cereal Sci. 21, 49–55. http://dx.doi.org/10.1016/S0733-5210(95)80007-7.p
ns, not significant.
[a]*E × C = elasticity × cohesiveness.*
[b]*E × C × S × N = elasticity × cohesiveness × softness × nonstickiness.*
Modified from Huang, S., Quail, K., Moss, R., Best, J., 1995. Objective methods for the quality assessment for northern style Chinese steamed bread. J. Cereal Sci. 21, 49–55. http://dx.doi.org/10.1016/S0733-5210(95)80007-7.

TABLE 9.10 Scoring System for the Assessment of Texture Properties of Southern-Style Steamed Bread

Quality Parameter	Full Score	TPA Parameter	Measurement	Score
Softness and cohesiveness	10	P1[a] (N)	<5.00	6
			5.01–6.50	7
			6.51–8.00	8
			8.01–9.50	9
			9.51–11.0	10
			11.01–12.5	9
			12.51–14.00	8
			14.01–15.50	7
			15.51–17.00	6
			17.01–18.50	5
			18.51–20.00	4
			20.01–21.50	3
			>21.51	2
Elasticity	10	A2/A1[b] (%)	>79.0	10
			78.9–78.0	9
			77.9–77.0	8
			76.9–76.0	7
			75.9–75.0	6
			74.9–74.0	5
			73.9–73.0	4
			72.9–72.0	3
			71.9–71.0	2
			<71.0	1
Stickiness	10	A3/A1[c] (%)	<1.50	10
			1.51–2.00	9
			2.01–2.50	8
			2.51–3.00	7

TABLE 9.10 Scoring System for the Assessment of Texture Properties of Southern-Style Steamed Bread—cont'd

Quality Parameter	Full Score	TPA Parameter	Measurement	Score
			3.01–3.50	4
			3.51–4.00	2
			>4.00	0

See Fig. 9.2 for sample texture profile analysis (TPA) graph.
[a]*P1 = peak force first compression cycle.*
[b]*A2/A1 = ratio of peak force area second compression cycle to the peak force area first compression cycle.*
[c]*A3/A1 = ratio of the negative force area during the first compression cycle to the peak force area during the first compression cycle.*
Reproduced with permission from Huang, S., Quail, K., Moss, R., 1998. The optimisation of a laboratory processing procedure for southern style Chinese steamed bread. Int. J. Food Sci. Tech. 33, 345–357. http://dx.doi.org/10.1046/j.1365-2621.1998.00166.

Similar TPA measurements can be used for Guangdong-style steamed bread.

Besides flour and wheat quality evaluations and use in breeding programs, some other reported applications of texture analysis include comparison of differences in texture due to changes in ingredients (Sze-Yin and Lai-Hoong, 2013), different addition rates (Wu et al., 2012b), or flour quality/processing interactions (Kondakci et al., 2015).

9.5.1.1 Guidelines for Use of Texture Analysis in Evaluation of Steamed Bread Quality

Slices from the steamed bread need to be cut to a standard thickness before texture analysis and the edges trimmed squarely. This is best achieved by use of an electric knife and a custom template or miter box to give a flat level surface and consistent dimensions. Huang et al. (1995) used a 28.5-mm slice. Most authors used a slice thickness between 20 and 30 mm, but this will depend on the size of the steamed bread.

Instrument settings are not prescriptive, but changes will affect the results obtained. For example, the different test speeds will not affect the peak height, but will affect the area. Selection of the correct load cell is also important. A typical TPA test setting was used by Sha et al. (2007): load cell 5 kg; pretest speed, 3 mm/s; test speed, 1 mm/s; posttest speed, 3 mm/s; and compression, 40%.

Puncture probes, knife blades, and shear cells are not suitable for use in TPA or SR evaluation of steamed bread. Instead, a cylinder probe should be used of sufficient size to compress and not tear the crumb of the cut steamed bread section.

Because of the variety of test profiles, probe diameters, and sample thicknesses used, each laboratory will need to calibrate instrumental analysis against sensory analysis to identify the range in sensory parameters and set up their own internal scoring system to convert instrumental results to a score.

The TA.XT software includes a macro for TPA which measures the following parameters: hardness, fracturability, cohesiveness, springiness, gumminess, chewiness, and resilience. These parameters do not have the same definitions as the terms used in sensory evaluation of steamed bread. The TPA parameter gumminess is applicable only to semisolids, and hence not steamed bread. Of the remaining TPA parameters, only hardness and resilience have been found to correlate significantly with sensory scores of hardness and springiness of steamed bread (Qian, 2005). These same TPA parameters correlated significantly with SR as measured on an SR cycle described earlier in this section. These correlations are shown in Table 9.11.

Further guidance and recommended instrument settings for TPA can be found at www.texturetechnologies.com.

9.5.2 Color Measurement

Portable tristimulus color meters such as those manufactured by Konica Minolta (http://www.konicaminolta.eu) have now become universally accepted for measurement of food color. They are widely used for flour, bread, and noodle color measurement. Color meters operate according to the Commission Internationale

TABLE 9.11 Correlation Coefficients for Relationship Between Sensory and Instrumental Texture Analysis

	Springiness Score (Sensory)	Hardness (TPA)	Resilience (TPA)	Stress Relaxation (SR)
Hardness score (sensory)	0.991**	−0.974**	0.974**	−0.983**
Springiness score (sensory)		−0.991**	0.991**	−0.995**
Hardness (TPA)			−0.992**	0.982**
Resilience (TPA)				−0.982**

TPA, texture profile analysis.
**p*<.05; ***p*<.01.
SR is defined by Huang, S., Quail, K., Moss, R., Best, J., 1995. Objective methods for the quality assessment for northern style Chinese steamed bread. J. Cereal Sci. 21, 49–55. http://dx.doi.org/10.1016/S0733-5210(95)80007-7.
Data from Qian, P., 2005. Effect of Wheat Flour Quality on Chinese Steamed Bread Stale and Research on Chinese Steamed Bread Anti-Staling (Ph.D. thesis). China Agricultural University, Beijing, China (In Chinese, English abstract).

de l'Eclairage color space parameters, with the most widely used systems being Yxy and L*a*b*. Steamed bread color assessment using the Minolta color meter is measured on the skin and on the crumb (Lin et al., 1990; Kruger et al., 1992). Huang et al. (1995) compared steamed bread color assessed visually by a trained sensory panel with results obtained instrumentally as Minolta Y,x,y and L*,a*,b* (Table 9.12). Minolta x and L*-b* index gave a higher correlation with sensory skin and crumb color than Minolta L* alone and the results had a high level of reproducibility. These findings indicate that tristimulus color measurements can be used to measure skin and crumb color. If required, a scoring system based on Minolta results can readily be set up in-house. There is a positive correlation between color as determined by the Chinese standard GB/T 17320-1998 and Minolta (Huang et al., 1995) skin color ($r=0.76^{**}$) and crumb color ($r=0.56^{**}$) (Chen et al., 2010).

9.5.3 Laser Technology

Laser-based scanning instrumentation for rapid, nondestructive, and reproducible determination of volume and the external parameters of bread, bakery, and other products have become available. Different models are available for different size products. The sample is positioned centrally in the instrument and then either rotated around the vertical axis or scanned to produce the 3-D scanned image. Data can be captured for later use and comparative purposes. Parameters are then derived for volume, height, length, width, weight, and specific volume. These instruments have much potential and have been used to measure steamed bread parameters (Kondakci et al., 2015).

TABLE 9.12 Correlation Analysis of the Relationship Between Sensory Scores and Minolta Color Meter Measurements for Skin and Crumb Color of Northern-Style Steamed Bread

	Sensory Skin Color	Sensory Crumb Color
Minolta reading "x" on skin	−0.97***	−0.95***
Minolta reading "x" on crumb	−0.94**	−0.97***
Minolta reading "L*" on skin	0.91**	0.92**
Minolta reading "L*" on crumb	0.80*	0.89**
Minolta (L*-b*) value on skin	0.96***	0.95**
Minolta (L*-b*) value on crumb	0.90**	0.96***

*$p<.05$; **$p<.01$; ***$p<.001$.
Modified from Huang, S., Quail, K., Moss, R., Best, J., 1995. Objective methods for the quality assessment for northern style Chinese steamed bread. J. Cereal Sci. 21, 49–55. http://dx.doi.org/10.1016/S0733-5210(95)80007-7.

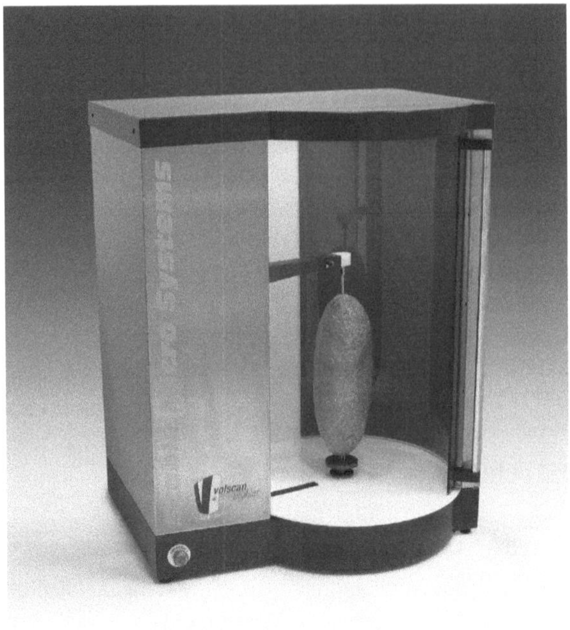

FIGURE 9.4 VolScan Profiler. *Courtesy Stable Micro Systems.*

Examples include the BVM (www.perten.com) and VolScan Profiler (www.stablemicrosystems.com) (see www.youtube.com/watch?v=9Hcs9EZhWbI). The VolScan Profiler is shown in Fig. 9.4.

9.5.4 Image Analysis

Image analysis provides an objective measurement tool for the evaluation of bread and baked product structure. High-resolution images of bread slices are captured and analyzed, providing measurements of crumb cell structure, distribution, color, and slice dimensions. One such dedicated instrument is the C-cell (www.c-cell.info). Limited applications have been reported for steamed bread, although significant correlations were reported with internal structure (Fang et al., 2013; He et al., 2007). Further work is required in this area and the technique would appear to offer another objective method of assessment.

9.6 CONCLUSION

There is a need for the standardization of laboratory methods for the preparation and assessment of steamed bread and for the adoption of objective methods in product evaluation. Instrumental methods will need to be validated and calibrated against current subjective methods if they are to replace them. It is possible to use instrumental methods of evaluation alone provided they have a meaningful relationship with the subjective methods they are intended to replace.

REFERENCES

Bourne, M.C., 1993. Texture measurements on finished baked goods. In: Kamel, B.S., Stauffer, C.E. (Eds.), Advances in Baking Technology. Blackie Academic & Professional, Glasgow, UK, pp. 135–151.

Chen, D.S., Zhang, Y., He, Z.H., Pena, R.J., 2010. Comparative study on evaluation methods for quality characteristics of northern style Chinese steamed bread. Sci. Agric. Sin. 43 (11), 2325–2333 (In Chinese, English abstract).

Chen, C.C., 2007. Taiwanese Style Steamed Bread Quality as Influenced by the Gluten Strength and Steam Generation Rate (M.Sc. thesis). University of Dayie, Taiwan (In Chinese, English abstract).

Chinese Standard Method GB/T 17320-1998, 1998. National Varieties for Wheat Quality for Specific End Uses, GB/T 17320-1998. Ministry of Agriculture, The People's Republic of China (In Chinese).

Chinese Standard Method GB/T 17320-2013, 2013. National Standard of the People's Republic of China, GB/T 17320-2013 in Replacement of GB/T 17320-1998, Quality Classification of Wheat Varieties. Released by the Grant Bureau of Quality Supervision, Inspection and Quarantine. The Chinese National Administration Committee for Standardisation, The People's Republic of China (In Chinese).

Chinese Standard Method SB/T 10139-93, 1993. Professional Standards for Specific End Uses of Wheat Flour for Steamed Bread, SB/T 10139-93. Ministry of Commerce, The People's Republic of China (In Chinese).

Fang, X., Sun, H., Cao, Y., Jiang, W., Chen, H., Yi, J., Qiu, Q., 2013. Study on the quality evaluation of steamed bread with image analyzer. J. Chin. Cereal Oils Assoc. 28 (6), 90–95 (In Chinese, English abstract).

Faridi, H.A., Rubenthaler, G.L., 1983. Laboratory method for producing Chinese steamed bread and effects of formula, steaming and storage on bread starch gelatinization and freshness. In: Proceedings 6th International Wheat Genetics Symposium, Kyoto, Japan, pp. 863–867.

Guo, B.L., Wei, Y.M., Zhang, G.Q., 2002. Study on quality evaluation methods of steamed bread. J. Cereal Crops 22 (3), 7–10 (In Chinese, English abstract).

He, S.M., Chen, D.S., Zhang, Y., He, Z.H., 2007. Analysis of cell structure of steamed bread by digital image analysis. Sci. Agric. Sin. 40 (1), 212–216 (In Chinese, English abstract).

Hou, G.G., Popper, L., 2007. Chinese steamed bread. In: Popper, L., Schäfer, S., Freund, W. (Eds.), Future of Flour – A Compendium of Flour Improvement. Verlag Agrimedia GmbH, Clenze, pp. 309–318. http://muehlenchemie.de/english/ (accessed 01.08.15.).

Huang, S., Quail, K., 1997. Effects of ingredients on the quality of Guangdong style steamed bread. In: Tarr, A.R., Ross, A.S., Wrigley, C.W. (Eds.), Cereals 97 Proceedings of the 47th Australian Cereal Chemistry Conference. Cereal Chemistry Division, Royal Australian Chemical Institute, Melbourne, Australia, pp. 81–84.

Huang, S., Betker, S., Quail, K., Moss, R., 1993. An optimised processing procedure by response surface methodology (RSM) for northern-style Chinese steamed bread. J. Cereal Sci. 18, 89–102. http://dx.doi.org/10.1006/jcrs.1993.1037.

Huang, S., Quail, K., Moss, R., Best, J., 1995. Objective methods for the quality assessment for northern style Chinese steamed bread. J. Cereal Sci. 21, 49–55. http://dx.doi.org/10.1016/S0733-5210(95)80007-7.

Huang, S., Quail, K., Moss, R., 1998. The optimisation of a laboratory processing procedure for southern style Chinese steamed bread. Int. J. Food Sci. Tech. 33, 345–357. http://dx.doi.org/10.1046/j.1365-2621.1998.00166.

Huang, S., Ruang, R., Quail, K., 2003. Eating quality and flour quality requirements of Guangdong style steamed bread. In: Black, C.K., Panozzo, J.F. (Eds.), Cereals 2003, Proceedings of the 53rd Australian Cereal Chemistry Conference. Cereal Chemistry Division, Royal Australian Chemical Institute, Melbourne, Australia, pp. 83–86.

Keeratipibul, S., Naphatrapi, L., Shinya, O., Shigeru, S., Yasushi, H., Somboon, T., 2013. Effects of the amount of Chinese steamed bun starter dough (CSB-SD) and the activation time on dough and bread properties. J. Food Process. Preserv. 37 (3), 232–244. http://dx.doi.org/10.1111/j.1745-4549.2011.00640.x.

Kim, C.S., Hwang, C.M., Song, Y.S., Kim, H.I., Chung, D.J., Han, J.H., 2001. Commercial wheat flour quality and bread making conditions for Korean-style steamed bread. J. Korean Soc. Food Nutr. 30, 1120–1128 (In Korean, English abstract).

Kondakci, T., Ang, A.M.Y., Zhou, W., 2015. Impact of sodium alginate and xanthan gum on the quality of steamed bread made from frozen dough. Cereal Chem. 92, 236–245. http://dx.doi.org/10.1094/CCHEM-03-14-0036-R.

Kruger, J.E., Morgan, B., Preston, K.R., Matsuo, R.R., 1992. Evaluation of some characteristics of Chinese steamed buns prepared from Canadian wheat flours. Can. J. Plant Sci. 72, 369–375.

Limley, H.A., Crosbie, G.B., Lim, K.K., Williams, H.G., Diepeveen, D.A., Solah, V.A., 2013. Wheat quality requirements for char siew bao made from Australian soft wheat. Cereal Chem. 90, 231–239. http://dx.doi.org/10.1094/CCHEM-11-12-0152-R.

Lin, Z., Miskelly, D.M., Moss, H.J., 1990. Suitability of various Australian wheats for Chinese-style steamed bread. J. Sci. Food Agric. 53, 203–213. http://dx.doi.org/10.1002/jsfa.2740530208.

Lin, T.C., 1983. Effects of Processing Methods and Wheat Flours on the Quality of Steamed Bread (M.Sc. thesis University of Kansas State, Kansas, USA. http://hdl.handle.net/2097/12974 (accessed 20.02.15.).

Liu, Z.D., Dong, C.W., Pang, K., 2006. Techniques on Flour Quality Improvement and Application. Chinese Light Industry Publisher, Beijing, China (In Chinese).

Liu, C.H., 2005. Production Techniques of Steamed Products. Chemical Industry Publisher, Beijing, China (In Chinese).

Qian, P., 2005. Effect of Wheat Flour Quality on Chinese Steamed Bread Stale and Research on Chinese Steamed Bread Anti-Staling (Ph.D. thesis). China Agricultural University, Beijing, China (In Chinese, English abstract).

Rubenthaler, G.L., Huang, M.L., Pomeranz, Y., 1990. Steamed bread. I. Chinese steamed bread formulation and interactions. Cereal Chem. 67, 471–475.

Sha, K., Qian, P., Wang, L.J., Lu, Z.H., Li, L.T., 2007. Effect of storage time on the physicochemical and sensory properties of man-tou (Chinese steamed bread). Int. J. Food Eng. 3 (3), 1556–3758. http://dx.doi.org/10.2202/1556-3758.1217. http://www.degruyter.com/view/j/ijfe (accessed 02.08.15.).

Su, D.M., 2005. Studies on Classification and Quality Evaluation of Staple Chinese Steamed Bread (Ph.D. thesis). Chinese Agricultural University, Beijing, China.

Sze-Yin, S., Lai-Hoong, C., 2013. Effects of maltodextrin and trehalose on the physical properties of Chinese steamed bread made from frozen doughs. Int. Food Res. J. 20 (4), 1529–1535.

Wang, L.K., Zhao, N.X., Cheng, A.H., Lan, J., 1998. A laboratory processing procedure and quality evaluation system for Chinese steamed bread. J. Chin. Cereals Oils Assoc. 13 (2), 29–32 (In Chinese).

Wu, C., Liu, R.S., Huang, W.N., Rayas-Duarte, P., Wang, F., Vao, V., 2012a. Effect of sourdough fermentation on the quality of Chinese northern-style steamed bread. J. Cereal Sci. 56, 127–133. http://dx.doi.org/10.1016/j.jcs.2012.03.007.

Wu, M.Y., Chang, Y.H., Shiau, S.Y., Chen, C.C., 2012b. Rheology of fiber-enriched steamed bread: stress relaxation and texture profile analysis. J. Food Drug Anal. 20, 133–142. http://www.fda.gov.tw/en/publishJFDAList.aspx (accessed 02.06.15.).

Xia, X.C., He, Z.H., Jin, L.H., Wang, L.H., Zhang, Y., Li, G.Y., 2013. Understanding the effects of HMW-GS and LMW-GS on processing quality using near-isogenic lines and development of functional markers for LMW-GS in wheat. In: He, Z.H., Wang, D. (Eds.), Proceedings 11th International Gluten Workshop China, August 12–15, 2012. D.F. International Maize and Wheat Improvement Center (CIMMYT), Mexico, pp. 107–112.

Zhang, P., He, Z., Zhang, Y., Xia, X., Chen, D., Zhang, Y., 2008. Association between % SDS-unextractable polymeric protein (%UPP) and end-use quality in Chinese bread wheat cultivars. Cereal Chem. 85, 696–700. http://dx.doi.org/10.1094/CCHEM-85-5-0696.

Zhu, F., 2014. Influence of ingredients and chemical components on the quality of Chinese steamed bread. Food Chem. 163, 154–162. http://dx.doi.org/10.1016/j.foodchem.2014.04.067 0308.

Chapter 10

Overcoming Steamed Bread Faults

10.1 INTRODUCTION

Steamed bread is more susceptible to quality faults than Western bread. Quality faults are closely related to the use of unsuitable ingredients and/or failure to control processing conditions. Because steamed bread has a soft and moist skin, problems associated with shape, appearance, and volume are very visually apparent.

10.2 APPEARANCE

A consumer will initially assess the quality of steamed bread visually by its color, smoothness, and shape. Steamed bread should have a shiny white color, smooth skin, and good size and regular shape as shown in Fig. 10.1.

10.2.1 Color

Steamed bread should have an attractive color, not too white and not too yellow. Factors affecting product color can include flour color and alkali addition, as well as some manufacturing parameters such as sheeting and steaming rate. Using blends of wheat flour with other flours to create product variety or nutritional benefits can result in a darker crumb color. Major color faults are discussed in the following sections.

10.2.1.1 Yellowish Color

In most cases, a high yellow color is due to overneutralization of dough with alkali after sourdough fermentation. Correct dough pH is in the range 6.4–6.7. Over pH 7, steamed bread has an alkali smell, bitter taste, and yellowish color. The yellow color is due to the naturally occurring flavonoids in the flour which are colorless below pH 7, but yellow at high pH. If the flour used in manufacture has high levels of yellow pigments (xanthophylls), the yellow color may be carried over into the steamed bread.

Steamed Breads. http://dx.doi.org/10.1016/B978-0-08-100715-0.00010-0

FIGURE 10.1 Steamed bread with very good appearance. *Courtesy Dr. H. Sun and Mr. Z. Yang.*

10.2.1.2 Extreme White Color

Steamed bread which is extremely white has usually been manufactured from flour which has been treated with bleaching agents such as benzoyl peroxide which destroy the natural flour yellow pigments.

10.2.1.3 Yellow Spots on the Surface and Crumb

Yellow spots on the surface and crumb of steamed bread can be caused by insufficient mixing after the alkali solution is added to neutralize the dough. Consumers will usually avoid purchase of product with this fault.

10.2.1.4 Dull Color

A dull dark skin and crumb can be due to a number of causes:

- A high extraction flour with high bran content will give steamed bread with dark skin and crumb color. A lower extraction flour will have lower ash, lower flour color grade, and higher Minolta L and give a better steamed bread color. A flour with less than a maximum ash content of 0.55% should be used in the manufacture of standard grade products, but for premium grade products the flour ash should be no more than 0.45% (as is basis).
- Use of rusty or discolored water.
- Underneutralization of the dough after sourdough fermentation.
- Too high relative humidity (RH) during proofing. The optimum is 60–90% RH, depending on season: in winter the RH in the first stage of proofing needs to be reduced; in summer the RH in the first stage of proofing needs increasing.
- Inadequate dough development. For handmade steamed bread, the dough pieces need to be properly kneaded before molding to avoid bread with a dull and dark color. Commercially, sheeting the dough before molding also helps to avoid the dull and dark color of the steamed bread.

10.2.1.5 Dull Dimpled Surface Perforated With Pin Holes

The surface of the steamed bread can become dimpled with small pin holes accompanied by a dull color due to too high proofing temperature and/or RH. Pinholes and dimpling can also occur if the steaming rate is too low, which effectively prolongs the proofing time. These problems can be overcome by increasing the amount of dusting flour in the late stage of machine molding, reducing the temperature and RH in proofing, and increasing the steaming rate (Liu, 2005).

10.2.2 Smoothness

Consumers prefer steamed bread with a smooth surface. Cracks, dimples, blisters, and brown spots often occur on the surface, making the product unappealing. These faults are due to poor flour quality and incorrect processing conditions.

10.2.2.1 Cracks

Cracks may occur on the surface of the product if the dough is too dry or the RH in the proofer is less than 60% or the expansion of the dough pieces is inhibited. This can be corrected by either increasing water addition or setting the proofer to 70–90% RH and increasing the proofing time. Cracks may also occur if the pH of the dough after neutralization is outside the optimum range of 6.4–6.7. Processing conditions that may cause cracking include undermixing of the dough, lack of control of dusting flour, and incorrect settings on the molding machine.

10.2.2.2 Dimples and Blisters

During the first 5 min in the steamer the dough expands quickly. Gas bubbles trapped immediately under the dough surface gradually grow in size. When the cover of the steamer is removed, the bubbles break, leaving blisters on the surface. Gas bubbles can largely be reduced in number and size if the dough is sheeted properly before molding, thus reducing or totally eliminating the surface blisters and dimples.

If the steaming rate is too high, blisters (Fig. 10.3), dimples, or brown spots (Fig 10.2 and 10.5) may appear on the surface of the product or partial collapse of the dough piece may occur (Fig. 10.4A).

Dimples and blisters can also be caused by the dough being too soft, undermixed, or underneutralized. High proofing temperatures, RH, and long proofing times can also be the cause of these defects.

10.2.2.3 Shrinkage

Steamed bread can shrink quickly once the steaming process is completed and the steamer is opened, even if they are smooth and round immediately beforehand (Fig. 10.2). The cause and extent of the shrinkage varies depending on conditions that will be discussed in this section. Shrinkage can occur when a

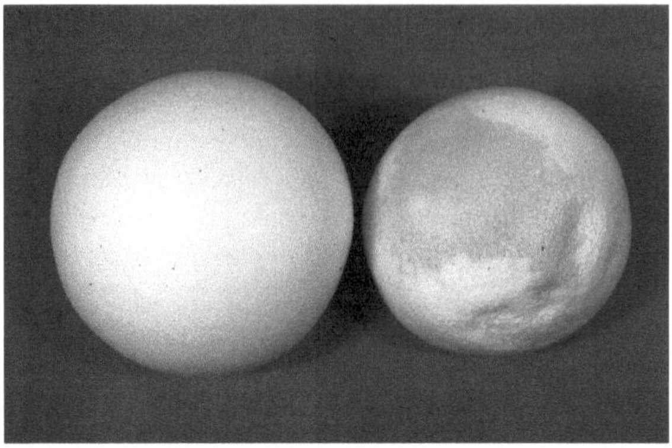

FIGURE 10.2 Two steamed breads: left, optimally mixed dough; right, dough was undermixed, causing shrinkage and brown spots.

FIGURE 10.3 Steamed buns with (left) and without (right) blisters.

flour with very strong dough properties and high protein content is used. In the most severe cases, the steamed bread shrinks to resemble a piece of rubber, where all the gas is expelled and the structure completely destroyed (Fig. 10.6). Shrinkage can also occur after resteaming or even during storage. This is a serious problem for steamed bread quality as the product is unsalable.

10.2.2.4 Causes of Shrinkage

Mixing steamed bread dough to the correct consistency is of critical importance for handling during processing, and for the best final quality. Dough mixed to its optimum consistency has a smooth, satiny, or "clear" appearance. A well-developed dough should be capable of being stretched to a uniform, semitransparent membrane before it ruptures. The developed gluten network can now retain gas

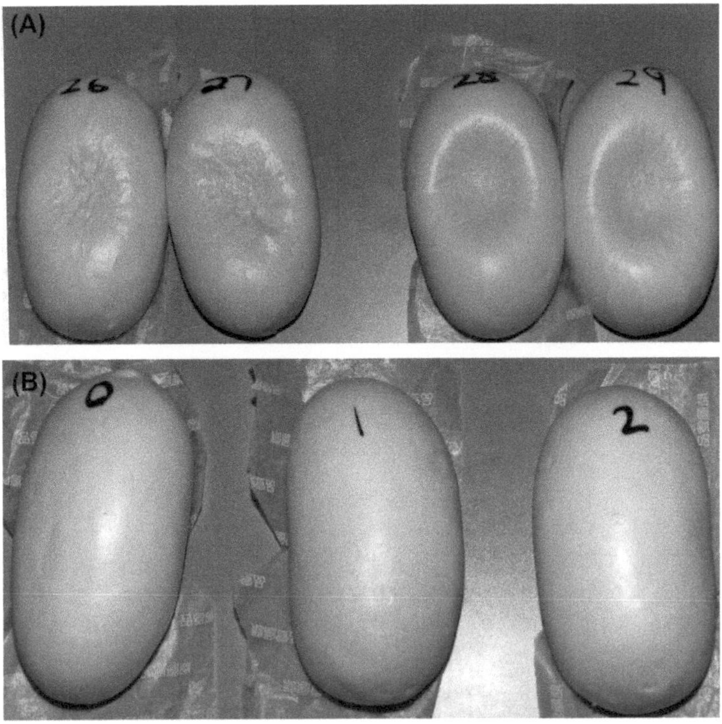

FIGURE 10.4 Steamed bread showing collapse (A) and normal steamed bread (B).

FIGURE 10.5 Partially shrunken steamed bread with jelly spots (brown color [gray in print version]) and dimples.

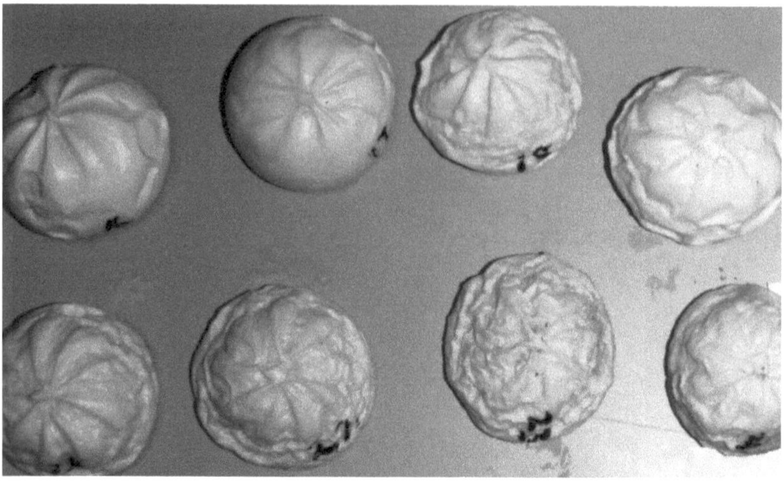

FIGURE 10.6 Steamed buns showing various levels of collapse.

bubbles when sudden changes in temperature and pressure occur such as when the steamer is opened. If the dough is undermixed, the gluten will be tough, less extensible, and not uniform. During steaming, gluten networks will be stretched as the dough piece expands. When the steamer is opened suddenly, the steamed bread cools down rapidly, and the internal pressure suddenly decreases. The resulting pressure differential together with the relaxation of the gluten network, now that no forces are stretching it, destroy the weak cell structure and gas is expelled, causing the shrinkage (Fig. 10.2, right). It was observed by one of us (Huang), using a transparent steamer cover, that when a long thin probe was inserted into the center of the steamed bread during steaming, there was less shrinkage when the steamer cover was removed due to lower pressure differential between the inside and outside of the bread.

Shrinkage sometimes occurs even in properly mixed doughs when overly strong flours are used. The shrinkage is caused by a very strong gluten network forming in the dough that relaxes after steaming.

Shrinkage may also occur when:

- the dough is underneutralized after sourdough fermentation,
- the dough pieces are overproofed,
- highly overfermented scrap dough is used,
- the steaming rate is too high or too low,
- the water is too hard, and
- the interior temperature and/or RH of the steamer is not uniform.

Shrinkage in steamed breads can be quite complex. Shrinkage can occur within batches or production runs, where some products shrink while the remainder do not. This demonstrates the high sensitivity of steamed bread quality to flour quality and processing conditions.

Bai (2010) systematically studied the causes of shrinkage of steamed bread. The number one cause of shrinkage was overproofing. Shrinkage was categorized as follows: severe shrinkage, medium shrinkage, light shrinkage, and partial collapse. Flours with very high dough strength caused severe shrinkage, whereas flours with very weak dough strength caused light shrinkage or partial collapse.

Yeast addition, sourdough addition, hardness of water, dough water addition, mixing time, proofing time, and steaming pressure all affected shrinkage. Changes in sourdough fermentation and proofing times can produce shrinkage. Extending the proofing time beyond optimum will result in a concomitant increase in shrinkage.

10.2.2.5 Prevention of Shrinkage

Flours with appropriate dough properties, neither too strong nor too weak, should be used as detailed in Table 3.6. Doughs should be mixed to optimum development, and the ingredient water used should be at the correct temperature so as not to inhibit yeast activity. Fermented doughs should be neutralized correctly and well mixed after alkali addition if sourdough fermentation is used. Doughs should be proofed for the correct time at the correct temperature and RH. The steamer should operate at the set temperature which should return to the set point quickly after the proofed dough pieces are placed into steamer.

Reworked or scrap dough (leftover dough from previous processing) should never be added to new mixes in the one-step process.

10.2.3 Weight and Dimensions

Steamed bread is produced in weights ranging from approximately 25–130 g. Dimensions are measured using the spread ratio (weight/height), but is only applicable with approximately spherical-shaped pieces. For northern style, spread ratio is around 1.2–1.5, and for southern style, 1.4–1.6.

10.2.3.1 Flattened Profile

Steamed breads should have a rounded profile. Finished product that has spread or collapsed is usually not salable. The flattened shape can be due to too much water added to the dough, or the dough being too soft, overmixed, or overproofed. Weak and rain-damaged flours may also cause the flattened shape (Fig. 10.7).

10.2.3.2 Uneven Shape

The molding machine should be checked and adjusted if necessary.

10.2.3.3 Small Volume

Small volume may be due to shrinkage, which can be corrected as described earlier. In the absence of shrinkage, dough water addition may be too low, flour dough strength may be too low, or yeast activity may be too low.

FIGURE 10.7 Steamed bread made from sound (A) and rain-damaged (B) wheat, showing a flattened profile typical of rain damage. *Reproduced with permission from Huang, S., 2014. Steamed bread. In: Zhou, W., Hui, Y.H., (Eds.), Bakery Products Science and Technology, second ed., Wiley Blackwell, Chichester, West Sussex, UK, pp. 539–562.*

10.3 FLAVOR

Normally steamed bread has a good wheaten flavor, slightly sweet after chewing. It may have a slight alkaline flavor if excess alkali has been used during processing.

Poor flavor may be due to:

- Wheat which is rain or frost damaged, moldy, insect damaged.
- Tainted ingredients due to poor storage conditions or exposure to external contaminants.
- Use of nonpotable water.

- Use of rancid or poorly deodorized vegetable oils employed to prevent product sticking to trays during steaming. Only food grade refined and deodorized vegetable oil should be used.
- Undesirable alkali flavor can be avoided by adjusting dough pH after fermentation to 6.4–6.7.

10.4 TEXTURE AND EATING QUALITY

10.4.1 Sticky Eating Quality

Steamed bread becomes sticky when:

- Flour is made from rain-damaged wheat.
- The flour has a high starch damage.
- The steaming rate is too low.
- The steaming time is too short.

10.4.2 Crumb Texture Is Too Firm

The causes of small volume and very firm crumb texture include:

- Doughs that are too tight. Water addition should be increased to optimum (not too high or a flattened shape may result).
- Insufficient yeast addition or low yeast activity. Increase yeast addition or fermentation time, or replace with a fresh batch of yeast.
- Short fermentation and proofing times. Increase fermentation and proofing time to optimum.
- Poor flour quality, generally low protein content and weak protein. Flour quality should be within the specifications detailed in Chapter 3, Table 3.6.

10.4.3 Steamed Bread Bottom Is Too Firm

This may be due to

- a too high steaming rate;
- prolonged time in the steamer;
- oversized dough pieces which were not processed correctly during molding;
- too much dusting flour on the dough piece.

10.4.4 Crumb Structure Is Too Coarse

Coarse crumb structure may be due to the use of weak flours and undermixed or overmixed doughs. Doughs should be mixed to optimum consistency, and not to the point of overdevelopment or breakdown. Sheeting is very beneficial for optimally or undermixed doughs to further develop the dough structure to achieve a fine and even crumb structure.

10.5 POTENTIAL OF FOOD ADDITIVES

Some of the enzymes and food additives discussed in Chapter 4 may have application in reducing a number of quality faults, especially where flour quality is less than ideal. According to Popper and Zhong (2015) a number of commercial products are available. The active ingredients include:

- Blends of lipase and hemicellulase which are reported to improve stability and extensibility of doughs and thus reduce product shrinkage.
- Lipase and emulsifiers which reduce yellowish color.
- Ascorbic acid addition to improve sticky eating quality.

10.6 INTEGRATING PRODUCT IMPROVEMENT INTO THE MANUFACTURING PROCESS

Manufacturers now place heavy emphasis on product quality and the reduction of waste and unsalable product. It is important to identify, document, and hence begin to control potential sources of product faults. Consistent product quality can be achieved by application of integrated quality and process control across the entire business entity, particularly in raw material specifications and purchase, product development, product manufacturing, warehousing, and sales.

A good quality control system will set the framework for systematic problem solving of quality faults by:

- Recording raw material supplier details and testing received raw materials to specification.
- Establishing and recording all production details and process parameters such as:
  - mixing time
  - water addition
  - proofer temperature, RH and time
  - steamer temperature, RH, pressure and time

  and ensuring their compliance in manufacture.
- Measuring and recording product parameters such as weight, specific volume, color, and defects and ensuring their compliance in manufacture.
- Using documented standard operating procedures.
- Recording all wastage such as scrap dough and other ingredients and the quantity of rejected or substandard product.

Techniques and data analysis tools for quality management can be used to identify causal factors for defective products. These systems operate with a team approach and involve manufacturing line operators as well as management and quality staff. Statistical analysis is used to enable fact-based decision making in solving quality faults and reducing the number of defective products. In a case study using the Six Sigma quality program (http://www.isixsigma.com)

applied to steamed bread production, the average defective shrinkage rate for custard steam buns was reduced by a remarkable 75% (Hung and Sung, 2011).

10.7 CONCLUSION

The increasing use of mechanized production for steamed bread has required manufacturers to closely control the selection, quality, and consistency of raw materials. This is coupled with the necessity for adequate process controls, and measurement of product quality. These tools will assist manufacturers to identify and then correct steamed bread faults leading to less wastage of product.

REFERENCES

Bai, J.M., 2010. Analysis of Phenomenon "Shrinkage" in Steamed Bread (M.Sc. thesis). Henan University of Technology, Zhengzhou, China (In Chinese, English abstract).

Huang, S., 2014. Steamed bread. In: Zhou, W., Hui, Y.H. (Eds.), Bakery Products Science and Technology, second ed. Wiley Blackwell, Chichester, West Sussex, UK, pp. 539–562.

Hung, H.S., Sung, M.H., 2011. Applying Six Sigma to manufacturing processes in the food industry to reduce quality cost. Sci. Res. Essays 6, 580–591. http://dx.doi.org/10.5897/SRE10.823. http://www.academicjournals.org/SRE (accessed 17.07.15.).

Liu, C.H., 2005. Production Techniques of Steamed Products. Chemical Industry Publisher, Beijing, China (In Chinese).

Popper, L., Zhong, X., April 2015. Chinese Steamed Bread. World Grain.

Chapter 11

Steamed Bread Spoilage, Food Safety, and Staling

11.1 INTRODUCTION

Traditionally, steamed bread is made and consumed within a day in summer and 2–3 days in winter. With recent changes in life styles, urbanization, availability of convenience foods, and expansion of large-scale production, the shelf life of steamed bread has taken on new importance.

11.2 STEAMED BREAD SPOILAGE

11.2.1 Characteristics of Steamed Bread Spoilage

When foods spoil, they become unacceptable or unsuitable for human consumption, which may be due to unacceptable changes in smell, flavor, appearance, or texture.

Visible signs of spoilage include off-flavors or -odors, growth of mold and yeast colonies on the surface of the bread, softening and stickiness of packaged bread, and hardening of the skin of unpackaged bread.

11.2.2 Microbial Spoilage

These visible signs are preceded by microbial contamination and growth. Microorganisms are distributed universally and their growth is affected by temperature, availability of nutrients, moisture, time, pH, and whether oxygen is present. Steamed bread, which has a moisture content between 35% and 45% and water activity (a_w) above 0.92 (Huang et al., 2003), has characteristics conducive to the growth of yeast, mold, and bacteria.

11.2.3 Marketplace Survey of Steamed Bread

In a marketplace survey of steamed bread purchased in Asian supermarkets in Sydney, Australia (Table 11.1), Huang et al. (2003) found that yeasts were the dominant colonies with moderate counts of 10^2–10^4 cfu/g. One sample had a high count of 3×10^6 cfu/g. Molds, which are a primary spoilage microorganism in baked goods,

Steamed Breads. http://dx.doi.org/10.1016/B978-0-08-100715-0.00011-2

TABLE 11.1 Range of Results for Standard Plate, Yeast and Mold Counts, Xerophilic Yeast and Mold Counts, and Coliforms for Steamed Bread Purchased in Sydney, Australia

Storage Condition	No. Tested	Standard Plate Count (cfu/g)	Yeast & Mold Count (cfu/g)[a]	Xerophilic Yeast and Mold Count (cfu/g)[a]	Coliforms (MPN/g)[b]
Room temperature	16	1.2×10^3 to $>3.0\times10^6$	<100 to $>3.0\times10^5$	<100 to $>3.0\times10^5$	<3–4
Refrigerated	6	3.0×10^2–1.7×10^5	<100–1.6×10^5	<100–1.5×10^5	<3–43
Frozen	6	<100–1.1×10^3	<100	<100	<3

[a] *Two spoiled refrigerated samples with visible molds are not included in these results.*
[b] *MPN, most probable number.*
Modified from Huang, Y., Jensen, N., Hocking, A., Miskelly, D., 2003. Microbiological Survey of Steamed Bread Products from Asian Grocery Stores in Sydney – Part 4. VAWCRC Report No. 25. http://ses.library.usyd.edu.au/handle/2123/2713 (accessed 30.06.14.).

were detected less frequently than yeasts. *Penicillium* spp. and *Cladosporium* spp. were detected. Standard plate counts were generally higher than desirable, with less than 10^5 regarded as an acceptable level for ready-to-eat food (NSW Food Authority, 2009). *Bacillus cereus* was detected in almost 50% of the samples. *B. cereus* is commonly detected in flour (Berghofer et al., 2003). It is a bacterium of concern as it produces spores and a toxin that can survive steaming and grow in the high a_w of steamed bread. Although coliforms were not detected in frozen samples at a detection limit of 3 most probable numbers per gram, they were detected in 14% of the samples stored at room and refrigerator temperatures. Lactic acid bacteria, which can cause off-odors, were found in samples stored at room temperature.

There was not much difference between products stored at room and refrigerator temperatures (Table 11.1). As expected, frozen products had lower counts across the range tested, indicating that freezing offers the best storage conditions to prevent spoilage.

11.3 FOOD SAFETY PROGRAMS

Food safety is a critical issue because it plays a significant role in the prevention of outbreaks of food-borne illness. It is becoming increasingly important with increasing consolidation and globalization of the food industry. Various systems and standards have developed since the 1950s and the best known of these is Hazard Analysis and Critical Control Points (HACCP). This is mandated in some countries. There are seven principles identified in a HACCP system (Codex Alimentarius Commision, 2009):

1. Conduct a hazard analysis.
2. Determine the critical control points (CCPs).
3. Establish critical limits for CCPs.
4. Establish a monitoring system for the control of the CCP.
5. Establish corrective actions when a CCP is not under control.
6. Establish verification procedures.
7. Establish documentation and record keeping appropriate to HACCP principles and their application.

Prerequisite programs that provide the right environment for HACCP include but are not limited to the following examples:

- a raw material supplier management program
- pest control
- equipment cleaning and maintenance
- recall procedures
- appropriate work place training

11.3.1 HACCP Plan

The HACCP plan must be specific to a product and processing line. The first step is to describe the product—physical description, specifications, distribution, ingredients, packaging, etc. Then a flow chart is set up and cross-checked against the physical process in the factory area. Hazards are identified for each step as:

- chemical hazards, eg, allergens such as nuts, soy, milk, and eggs; cleaning chemicals, pesticide residues;
- biological hazards, eg, pathogens present in raw materials, cross-contamination during manufacture, poor storage of cooked product at inappropriate temperature, toxins produced in the product, pathogens surviving the cooking and steaming process;
- physical hazards, eg, foreign material in raw materials such as glass, metal, wood, and plastic; poorly maintained equipment; insects and rodent contaminations.

Based on the hazard analysis, a risk assessment is carried out. CCPs are identified, together with a monitoring system and control measures. The process is verified and a controlled documentation system established. HACCP logos are often included on steamed product packaging.

11.3.2 Integration With Other Management Systems

It is becoming common for HACCP to be used as part of a larger food safety management system such as ISO 22000. This can be used by all organizations involved in the food chain, both national and global. The Global Food Safety Initiative provides international recognition of third-party certification to benchmark equivalence between the various food safety schemes throughout the global supply chain (www.mygfsi.com).

11.4 PRESERVATION AND SAFE STORAGE

As described earlier, microbial contamination and growth can be minimized by good control of potential biological hazards before, during, and after manufacture. Preservatives and choice of storage conditions may be used to retard microbial growth and hence extend shelf life.

11.4.1 Techniques to Prolong the Shelf Life of Steamed Bread

Preservatives such as sorbic or propionic acid and their sodium, potassium, or calcium salts; sodium diacetate; sodium benzoate; fumaric acid; and acetic acid (vinegar) can be used singly or in combination to inhibit or retard yeast and mold growth and hence extend the shelf life of breads including steamed bread. Their use and rate of usage are usually governed by food standards in the relevant country. Generally, the rate of addition is lower than permitted in the standards due to their retarding action on yeast added to the dough. Bread flavor may also be affected. The efficacy of these preservatives as mold inhibitors differs and is affected by temperature, pH, and water activity.

Although there have been extensive studies on the use of preservatives to extend the shelf life of baked products, there have been few reported applications for steamed bread. The combined use of sodium benzoate (0.08%) and calcium propionate (0.08%) was found to be effective for the storage of steamed bread for 48 h at 38°C without deterioration in quality (Liu, 2005).

11.4.2 Storage Conditions

Refrigerated storage can extend the shelf life of steamed products, although the rate of staling for bakery products is higher than at ambient temperatures. Frozen products can have a shelf life of up to 6 months. Modified atmosphere packaging (MAP) is widely used to extend the shelf life of baked products. Generally the inert gas used is 100% carbon dioxide, or a combination of carbon dioxide and nitrogen. A patent application (Zhang, 2008) claims a shelf life of 90 days for steamed bread packed at 70°C using MAP with carbon dioxide. Another method of extending shelf life is to use oxygen scavengers, which are sachet inserts of oxygen-absorbing material. The application of a combination of oxygen scavenger and an ethanol emitter was applied to steamed bread (Sheng et al., 2015) showing potential application in extending the shelf life and reducing microbial growth.

Apart from refrigerated storage, these storage and packaging methods have a high cost, and are suitable only for high-value products, or those traded internationally.

11.5 STALING

Bread staling is an important consideration in Western bread and baked products and is characterized by firming of the crumb and changes to the crust as moisture migrates from the crumb to the crust and organoleptic properties become unacceptable. Although the mechanism of bread staling is complex, it appears that starch retrogradation/recrystallization are part of the staling process (Gray and Bemiller, 2003). Opinions vary as to the role of amylose and amylopectin, although amylopectin may be more involved and whether protein and pentosans contribute to staling. Staling of Western bread has been studied for many decades, and attention is drawn to a review by Gray and Bemiller (2003) of the molecular basis and control of bread staling. The small body of work on steamed bread has largely built on the basis of mechanisms developed for staling of Western bread.

Although steamed bread has a shorter shelf life, the physical manifestations of staling are similar in that the crumb becomes firmer and tougher (Zhu, 2014). The sensory attributes of steamed bread, including softness, stickiness, cohesiveness, and elasticity decreased dramatically within 24h (Sha et al., 2007). This is illustrated in Fig. 11.1, which shows the increase in crumb firmness over 3 days.

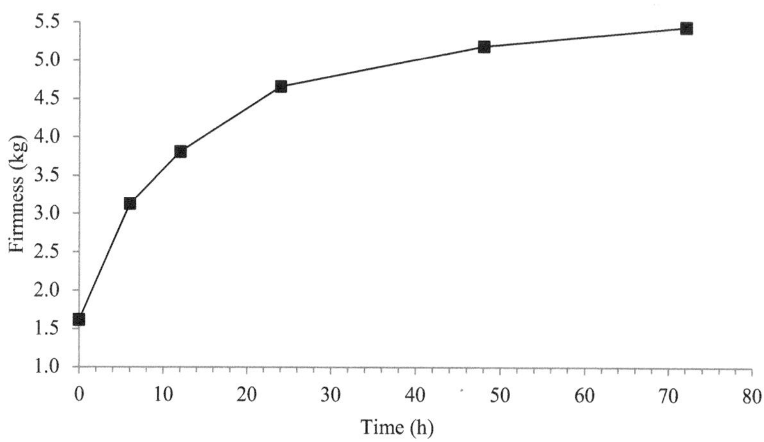

FIGURE 11.1 Increase in crumb firmness of northern-style steamed bread measured using texture analysis. *Data from Sha, K., Qian, P., Wang, L.J., Lu, Z.H., Li, L.T., 2007. Effect of storage time on the physicochemical and sensory properties of man-tou (Chinese steamed bread). Int. J. Food Eng. 3 (3), 1556–3758. http://dx.doi.org/10.2202/1556-3758.1217.* http://www.degruyter.com/view/j/ijfe *(accessed 02.08.15.).*

Various authors have postulated that amylose was responsible for the initial firming of steamed bread during storage (Qian, 2005; Zhu, 2007). By contrast, differential scanning calorimetry was used to show that amylopectin recrystallization was associated with firming (Sha et al., 2007). Both starch and protein have been shown to impact on the firmness of northern-style steamed bread during storage (Qian, 2005).

Western-style bread exhibits a moisture gradient within the loaf after baking, with about 40% moisture in the center crumb and about 15% in the crust. During storage of pup loaves (100 g flour and similar in size to steamed bread), the center crumb moisture decreases gradually and the crust moisture increases by about 5% in the first 24 h and more gradually thereafter (Czuchajowska and Pomeranz, 1989).

The pattern of moisture migration in steamed bread is different. Sha et al. (2007) reported that immediately after removing from the steamer, the skin moisture was 41.5% and the crumb moisture 39%. The skin moisture decreased rapidly as surface moisture evaporated, and then equilibrated to 39.5% on cooling to 22°C. Over the next 72 h, there was little difference between the skin, crumb close to the surface (intermediate zone), and center crumb, which ranged from 38% to 39.5%. This phenomenon is quite similar to Western bread stored without the crust, where there is little change in the crumb moisture and a_w with increasing storage time, but significant firming and amylopectin recrystallization (Baik and Chinachoti, 2000). This may help explain the more rapid staling rate of steamed bread compared with Western bread.

11.6 ADDITIVES TO EXTEND SHELF LIFE

As with Western bread, additives have been shown to extend the shelf life of steamed bread. Treatments with enzymes, emulsifiers, shortenings, hydrocolloids, and other additives have been applied to steamed bread, showing potential for longer shelf life (Table 11.2). These additives need to be cost-effective and maintain product shape, quality, and spread ratio to be acceptable. Commercial enzymes such as Novamyl Steam formulated for steamed bread are becoming available.

11.7 CONCLUSIONS

Consumers expect safe food no matter where that food is produced. The number and size of food manufacturing entities is growing as is the sale of steamed products into both the domestic and international markets. The past few decades have seen the development of food safety and quality standards and systems which are now essential to protect both manufacturers and consumers. Expectations of high quality have also increased, and there is now a need for manufacturers to extend the shelf life of steamed products using all available food ingredients and technologies.

TABLE 11.2 Additives for Extending Shelf Life of Steamed Bread

Treatment	Optimum Level	Effect or Improvement	References
Novamyl® Steam	50 mg/kg	Maintains softness during storage up to 3 days	Xu et al. (2014)
Max-life© amylase	300 mg	Reduces firmness during 3-day storage	Qian (2005)
Grindamyl© A1000 amylase	100 mg		
Maltogenic α-amylase	50 mg	Maltogenic α-amylase better antistaling agent than fungal α-amylase	Wang et al. (2006)
Fungal α-amylase	30 mg		
SSL, CSL, and GML	0.6%	Reduce firmness during 3-day storage	Qian (2005)
Trehalose	3%	Reduces firmness during 2-day storage	Hu et al. (2012)
Shortenings including lard, shortening, butter	6%	Retard staling. Did not affect firming during first few hours of storage	Li and Chen (1999)
			Qian (2005)
			Wang et al. (2006)
Sodium alginate	0.20%	Softer and more resistant to staling. Low spread ratio and specific volume	Sim et al. (2011)
Konjac glucomannan	0.80%		
Low temperature defatted soybean hull	3%	Soybean hull is better than soybean flour in retarding staling	Xiao et al. (2005)
Soybean flour	2%		
Soy flour	0.60%	Staling rate was reduced by 35%	He et al. (2009)
Glycerol plus lactic acid	2.50%+ 0.25%	Lower water activity Longer shelf life (8 days)	Laohasongkram et al. (2011)

CSL, calcium stearoyl-2-lactylate; GML, glyceryl monolaurate; SSL, sodium stearoyl-2-lactylate.

REFERENCES

Baik, M.Y., Chinachoti, P., 2000. Moisture redistribution and phase transitions during bread staling. Cereal Chem. 77, 484–488.

Berghofer, L.K., Hocking, A.D., Miskelly, D., Jansson, E.J., 2003. Microbiology of wheat and flour milling in Australia. Int. J. Food Microbiol. 85, 137–149. http://dx.doi.org/10.1016/S0168-1605(02)00507-X.

Codex Alimentarius Commision (CAC), 2009. Basics Text on Food Hygiene, fourth ed. Food and Agricultural Organization of the United Nations, Rome, pp. 23–33.

Czuchajowska, Z., Pomeranz, Y., 1989. Differential scanning calorimetry, water activity and moisture contents in crumb center and near-crust zones of bread during storage. Cereal Chem. 66, 305–309.

Gray, J.A., Bemiller, J.N., 2003. Bread staling: molecular basis and control. Compr. Rev. Food Sci. Food Saf. 2 (1), 1–21. http://dx.doi.org/10.1111/j.1541-4337.2003.tb00011.x.

He, Y.Q., Ma, Y.D., Li, J.B., 2009. Effects of soybean phosphate flour on quality and staling of steamed bread. J. Agric. Sci. Technol. 3 (11), 33 (In Chinese).

Hu, Y., Li, X., Zhou, Q., 2012. Effect of trehalose on anti-aging of steamed bread. In: Zhang, T.C., Ouyang, P., Kaplan, S., Skarnes, B. (Eds.). Proceedings of the 2012 International Conference on Applied Biotechnology (ICAB 2012), vol. 1. Springer-Verlag, Heidelberg, pp. 547–555. http://dx.doi.org/10.1007/978-3-642-37925-3_151.

Huang, Y., Jensen, N., Hocking, A., Miskelly, D., 2003. Microbiological Survey of Steamed Bread Products from Asian Grocery Stores in Sydney – Part 4. VAWCRC Report No. 25 http://ses.library.usyd.edu.au/handle/2123/2713 (accessed 30.06.14.).

Laohasongkram, K., Poonnakasem, N., Chaiwanichsiri, S., 2011. Process development of shelf-stable Chinese steamed bun. J. Food Process. Eng. 34, 1114–1124.

Li, L.T., Chen, X.Y., 1999. The improvement of steamed bread quality by the addition of lard. J. Chin. Cereals Oils Assoc. 14 (2), 47–50 (In Chinese, English abstract).

Liu, C.H., 2005. Production Techniques of Steamed Products. Chemical Industry Publisher, Beijing, China (In Chinese).

NSW Food Authority, 2009. Microbiological Quality Guide for Ready-to-Eat Foods. http://www.foodauthority.nsw.gov.au/documents/science/microbiological_quality_guide_for_rte_food.pdf (accessed 02.06.15.).

Qian, P., 2005. Effect of Wheat Flour Quality on Chinese Steamed Bread Stale and Research on Chinese Steamed Bread Anti-Staling (Ph.D. thesis). China Agricultural University, Beijing, China (In Chinese, English abstract).

Sha, K., Qian, P., Wang, L.J., Lu, Z.H., Li, L.T., 2007. Effect of storage time on the physicochemical and sensory properties of man-tou (Chinese steamed bread). Int. J. Food Eng. 3 (3), 1556–3758. http://dx.doi.org/10.2202/1556-3758.1217. http://www.degruyter.com/view/j/ijfe (accessed 02.08.15.).

Sheng, Q., Guo, X.N., Zhu, K.X., 2015. The effect of active packaging on microbial stability and quality of Chinese steamed bread. Packag. Technol. Sci. 28, 775–787. http://dx.doi.org/10.1002/pts.2138.

Sim, S.Y., Noor Aziah, A.A., Cheng, L.H., 2011. Characteristics of wheat dough and Chinese steamed bread added with sodium alginates or konjac glucomannan. Food Hydrocoll. 25, 951–957. http://dx.doi.org/10.1016/j.foodhyd.2010.09.009.

Wang, X., Yang, H., Yao, J., Shen, J., 2006. Comparison of effect of fungal α-amylase and maltose α-amylase on the improvement of steamed bread quality. J. Food Sci. Technol. 10, 48–52 (In Chinese with English abstract).

Xiao, A., He, D., Zhang, S., 2005. Promoting quality and postponing staling of steamed bread by adding baked soybean flour or dietary fiber from soybean hulls. J. Chin. Cereal Oils Assoc. 20 (4), 73–76 (In Chinese, English abstract).

Xu, Q.L., Sato, K., Yamaguchi, S., 2014. Steamed Bun Quality Improving Agent and Use Thereof. US Patent application US 2014/0037790 http://www.google.com/patents/US20140037790 (accessed online 13.08.15.).

Zhang, B., 2008. Fresh-Keeping Storing Method for Steamed Bun Products. Patent CN101124923 (In Chinese, English abstract) http://worldwide.espacenet.com (accessed online 03.08.15.).

Zhu, F., 2007. Study on the Quality of the Physicochemical Properties of Wheat Flour and Their Relationship with the Quality of the Chinese Steamed Bread (M.Sc. thesis). Wuhan Polytechnic University, Wuhan, China (In Chinese, English abstract).

Zhu, F., 2014. Influence of ingredients and chemical components on the quality of Chinese steamed bread. Food Chem. 163, 154–162. http://dx.doi.org/10.1016/j.foodchem.2014.04.067 0308.

Chapter 12

Nutrition of Steamed Products

12.1 INTRODUCTION

As described in Chapter 1, steamed bread is the staple food in the wheat-growing areas of northern China and rice is the staple food of the rice-growing areas of southern China. In northern China, steamed bread may be consumed at all meals in combination with other foods to provide a balanced diet but in the south it is eaten mostly at breakfast or as a snack. In most other parts of Asia, where steamed buns are common, they are consumed at breakfast or as a snack at any time.

12.2 NUTRITIONAL COMPOSITION OF STEAMED PRODUCTS

Breads and cereal-based foods are an important source of energy, carbohydrate, protein, B group vitamins (especially thiamine, niacin, and riboflavin), and minerals (calcium, phosphorus, and iron) in the diet. Table 12.1 compares the nutritional composition of commonly consumed wheat products including steamed bread and buns. The staple traditional northern-style Chinese steamed bread has no added fat or salt, and thus has the lowest fat and salt content compared with Australian bread and instant noodles that are a popular snack throughout Asia. If used, salt is generally added at no more than about 0.5% of the total flour weight, accounting for the low sodium levels in the finished bread. Thus northern-style steamed bread is suitable for those wishing to decrease salt intake in their diet. Lower sodium intake is associated with a decrease in blood pressure and a reduced risk of stroke and coronary heart disease as shown in meta-analyses (Aburto et al., 2013).

Steamed buns (Table 12.1) may provide an additional contribution of fat and sugar derived from the filling as well as the encasing bread. Plain Guangdong-style steamed breads also have added sugar and fat (Table 12.2), but a low salt content and although they vary greatly in their fat and sugar contents, the high amounts are added for consumer acceptance but are not required for functionality in processing (see Sections 4.4 and 4.5). Reduction of fat and sugar levels by manufacturers would contribute to more healthy products in both the domestic and international markets where they are readily available.

Steamed Breads. http://dx.doi.org/10.1016/B978-0-08-100715-0.00012-4

TABLE 12.1 Nutrient Composition of Various Breads and Instant Noodles Made From Wheat Flour

	White Bread Australia	Wholemeal Bread Australia	Instant Noodles (Cooked, Drained of Excess Water)	Steamed Bread Standard Flour China	Steamed Bread Special Refined Flour China	Vegetable Steamed Bun Singapore	Red Bean Paste Bun Singapore
Source	**NUTTAB 2010**	**NUTTAB 2010**	**NUTTAB 2010**	**China Food Nutrition Network**	**China Food Nutrition Network**	**Singapore Health Promotion Board**	**Singapore Health Promotion Board**
Moisture (%)	38.5	40.4	61.3	40.5	47.3	49.6	27.5
Protein (%)	9.7	9.0	3.9	7.8	6.2	5.4	6.0
Fat (%)	2.1	2.9	8.0	1.0	1.2	6.1	9.5
Carbohydrate (%)	45.0	39.7	22.8	49.8	44.2	29.7	49.2
Dietary fiber (%)	2.8	6.3	3.5			3.3	3.2
Sugars (%)	3.5	3.0	0.0			5.0	22.7
Thiamine (mg/100 g)	0.398[a]	0.51[a]	0.125			0.02	0.07
Riboflavin (mg/100 g)	0.049	0.085	0.145			0.14	0.16
Sodium (mg/100 g)	456	468	291	165	165	365	124
Energy (kJ/100 g)	1027	982	778	988	879	816	1281

[a]*Flour fortified with thiamine.*

TABLE 12.2 Composition of Guangdong-Style Steamed Bread Purchased From Supermarkets in Sydney

	Taiwan	Taiwan	West Malaysia	West Malaysia	Malaysia	Singapore	Singapore	Shenzhen	Hong Kong
Moisture (%)	35.6	36.9	30.6	28.9	37	30.6	32.3	33.6	32.2
Protein (%)	7.6	7.6	5.9	5.9	5.7	5.6	5.9	5.1	4.9
Fat (%)	3.0	2.1	7.0	7.5	1.4	8.2	8.3	1.8	5.2
Sugar (%)	7.7	2.1	10.1	9.7	19.1	4.0	4.0	9.1	12.1
Sodium (mg/100g)	115	28	155	156	85	124	121	39	79

12.3 FLOUR FORTIFICATION

It has been recognized by WHO and UNICEF that supplementation of industrially milled wheat flour and maize products with B group vitamins, minerals, and vitamins A and D has a positive impact on public health. Fortification at the country and international level is assisted by an international organization, the Food Fortification Initiative (www.ffinetwork.org). To date, 83 countries have legislated to fortify wheat flour and most include folic acid to prevent neural tube birth defects and iron to prevent anemia. The Asian countries which have mandated flour fortification are Indonesia (iron, folic acid, and zinc) and the Philippines (iron and vitamin A), with rice fortification planned in China, Cambodia, and Indonesia.

A study of the impact on the quality of various nutrients (blends of iron from different chemical sources, folic acid, vitamin B_{12}, vitamin A, thiamine, and riboflavin) added to steamed breads in China, Indonesia, and the Philippines showed:

- slightly darker crumb and in some instances grayish spots depending on the chemical source of iron;
- no effect on skin appearance or crumb texture;
- no effect on taste, mouthfeel, or aroma.

The results show that acceptable steamed products can be obtained from fortified flour and detail the retention rates of the added nutrients. Further details of this study can be obtained at www.ffinetwork.org or in a report by van den Wijngaart and Codling (2013).

12.4 GLYCEMIC INDEX

Glycemic index (GI) is a measure of the impact on blood glucose of digestible carbohydrates of foods (www.glycemicindex.com). Foods are classified as high (GI ≥ 70), moderate (GI 56–69), or low GI (GI ≤ 55). Foods with a high GI are rapidly digested and blood glucose rises rapidly after consumption. Low-GI foods in which starch is digested slowly show benefits in treatment of diabetes, weight control, improvement of satiety, and prevention of cardiovascular disease. As the rates of type 2 diabetes and obesity are increasing rapidly around the world, and particularly in Asia, there is considerable interest in GI of foods in many countries. The rate of diabetes in the adult population of China in 2010 was 11.6%, compared with 1% in 1980 (Xu et al., 2013) and the prevalence of obesity in male adult residents of Beijing in 2011 was 20.3% prompting a public health campaign for weight reduction.

Like white Western bread, steamed bread is a high-GI food. Unlike Western bread, there is little published information on the GI of steamed bread. Lau et al. (2015) reported that the GI of white Western bread was 71 ± 5, whereas that of steamed bread was 68 ± 4, which was not significantly different. Other workers have reported higher GI for steamed bread in the range 88.1–93.1 (Su-Que et al., 2013; Yang et al., 2006), which may be due to different methodologies, ingredients, or methods of preparation.

Many ingredients are used by the food industry in the West to lower the GI of baked products. These have potential for use in Chinese steamed breads, but have largely remained unexploited. Examples include guar gum, polyols, β-glucan-rich barley and oat fiber, arabinoxylans, and intact grains (Fardet et al., 2006). Replacement of conventional wheat flour with flour from high-amylose wheats gave steamed breads with significantly lower GI (Su-Que et al., 2013), confirming previous studies with baked bread.

Other factors such as milling, particle size, and processing conditions can be utilized to potentially lower GI. A study by Lau et al. (2015) demonstrated that bread processing conditions—mixing time and intensity, proof time, and method of cooking (whether baked or steamed)—could be used to manipulate GI, thus offering more opportunities for steamed bread manufacturers to develop low-GI products.

12.5 WHOLEGRAIN

Most Western countries now recommend wholegrain cereal foods in preference to refined flours. Consumption of foods based on wholegrain offers public health benefits in terms of reduction of risk of heart disease, stroke, birth defects, some cancers, and diabetes. There has been a trend in many countries in the West toward increased consumption of wholegrain and multigrain breads as well as the inclusion of whole seeds such as linseed and cracked or roughly milled grains such as soy, wheat barley, etc.

The accepted definition of wholegrain is that all parts of the grain—endosperm, bran, and germ—should be present in the proportions they were present in the original seed (www.aaccnet.org; www.healthgrain.org). This is mandated in some countries, where mixtures of bran and white flour do not fulfill their requirements for labeling as wholemeal or wholegrain.

Wholegrain flour is produced at the end of the roller milling process by combining all of the flour, bran, and pollard streams with the germ. Traditional style stone-ground grain satisfies the definition of wholegrain flour provided no fractions have been removed. Wholegrain or wholemeal flour contains more dietary fiber, phytochemicals, fermentable carbohydrates, B group vitamins, and minerals than refined flour, as many of these nutrients are concentrated in the bran layers and the germ. Unless the bran and germ fractions are heat treated, or kept refrigerated, wholemeal flour has a shorter shelf life than white flour due to the increased levels of free fatty acids which are more prone to the development of oxidative rancidity.

Although there are steamed products in the market which claim to be wholegrain, most do not contain 100% wholemeal. Much product development needs to be done to integrate wholegrain and wholemeal into steamed bread, offering increased variety of products to the consumer in addition to health benefits. The outcomes of the EU project HEALTHGRAIN (www.healthgrain.org) identified many developmental pathways for new technologies and identifying, promoting, and commercializing the health potential of whole grains.

12.6 ANTIOXIDANTS IN GRAIN

Antioxidants are naturally occurring bioactive compounds in grain that may protect against oxidative damage to DNA, proteins, and lipids in human cell membranes. The antioxidants found in wheat occur mainly in the germ, bran, and aleurone layers and include flavonoids, phenolics, carotenoids, tocopherols, and lignin. Some carotenoids such as lutein may have additional benefits in eye health. There is considerable variation in the antioxidant content of wheat varieties even within white and red wheat milling classes, but colored wheats such as black and purple have been reported to have higher levels of anthocyanins (water-soluble pigments) and higher levels of phenolics. The degree of refinement during milling is indicative of the retention of the antioxidant content, with whole wheat flour having a significantly higher phenolic and flavonoid content than partially debranned and white flour (Li et al., 2015). Additional processing into food decreases phenolic and flavonoid content, with steamed bread reported by the same authors as containing only 15–20% of that found in the original flour.

Further work is required to investigate the antioxidant composition of different wheats and also different environmental factors. This would enable breeding programs to select wheat lines with antioxidant properties. Work is also needed to investigate the potential for different processing methods and ingredients to increase the retention of antioxidants during manufacture of flour and end products.

12.7 FOOD AS MEDICINE

The concepts of food, nutrition, and medicine are much more intertwined in Chinese culture than in the West. The term *yao shi tong yuan* has a long history in China and implies that food and medicine are derived from the same source. Traditional Chinese cuisine is very varied and includes pickled vegetables, salted foods such as jellyfish, fermented foods such as soy sauce, and tofu extracted from soybeans. It is not surprising that supplementation of steamed bread with health promoting ingredients has received attention.

12.7.1 Supplementation With Green Tea

Green tea contains high levels of catechins. Catechins are flavonoids and are a source of polyphenols which are natural antioxidants. A number of authors have suggested addition of green tea extract to steamed bread (Ananingsih et al.,

2013; Goh et al., 2015; Oh et al., 2002) at addition rates between 0.5% and 3.2%. Steamed bread with added green tea extract was darker in color and had a lower specific volume, but the process could be optimized or fungal α-amylase added to obtain acceptable consumer products. Average retention rates in the finished product were in excess of 80%. Green tea catechins inhibit α-amylase and α-glucosidase enzymes involved in starch digestion and have a potential role in lowering glycemic response (Goh et al., 2015), but this will need to be confirmed with in vivo testing.

12.7.2 Supplementation With Mushrooms and Other Fungi

Mushrooms and other fungi can be a source of functional ingredients such as polysaccharides and fungal chitins and are used in Chinese medicine. They are reported to act as antiinflammatory and anticancer agents. Addition rates of up to 5% have been reported for Enokitake (winter or golden mushroom), Shiitake stipe (forest mushrooms), and silver ear fungus (Tsai et al., 2010; Tseng et al., 2011; Yen et al., 2015). As with many other supplemented products, they have darker color and a smaller specific volume and this can be tolerated by consumers.

12.8 OTHER INGREDIENTS

Other ingredients with perceived health benefits such as blended cereal flours, dietary fiber hulls, and resistant starch are covered in Chapter 4.

12.9 CONCLUSION

Steamed bread is an ideal carrier for the addition of macroingredients and microingredients, and there is the potential for development of innovative healthy products. This has already occurred in many Western countries, where there is increased demand for different varieties of bread in the market.

REFERENCES

Aburto, N.J., Ziolkovska, A., Hooper, L., Elliott, P., Cappuccio, F.P., Meerpohl, J.J., 2013. Effect of lower sodium intake on health: systematic review and meta-analyses. Brit. Med. J. 346, f1326. http://dx.doi.org/10.1136/bmj.f1326.

Ananingsih, V.K., Gao, J., Zhou, W., 2013. Impact of green tea extract and fungal alpha-amylase on dough proofing and steaming. Food Bioprocess Technol. 6, 3400–3411. http://dx.doi.org/10.1007/s11947-012-0986-3.

Fardet, A., Leenhardt, F., Lioger, D., Scalbert, A., Rémésy, C., 2006. Parameters controlling the glycaemic response to breads. Nutr. Res. Rev. 19, 18–25.

Goh, R., Gao, J., Ananingsih, V.K., Ranawana, V., Henry, C.J., Zhou, W., 2015. Green tea catechins reduced the glycaemic potential of bread: an in vitro digestibility study. Food Chem. 180, 203–210. http://dx.doi.org/10.1016/j.foodchem.2015.02.054.

Lau, E., Soong, Y.Y., Zhou, W., Henry, J., 2015. Can bread processing conditions alter glycaemic response? Food Chem. 173, 250–256. http://dx.doi.org/10.1016/j.foodchem.2014.10.040.

Li, Y., Ma, D., Sun, D., Wang, C., Zhang, J., Xie, Y., Guo, T., 2015. Total phenolic, flavonoid content, and antioxidant activity of flour, noodles, and steamed bread made from different colored wheat grains by three milling methods. Crop J. 3, 328–334. http://dx.doi.org/10.1016/j.cj.2015.04.004.

Oh, Y.K., Kim, C.S., Chang, D.J., 2002. Optimization of steamed bread making with addition of green tea powder using response surface methodology. J. Korean Soc. Food Sci. Nutr. 31, 451–459 (In Korean with English abstract).

Su-Que, L., Ya-Ning, M., Xing-Pu, L., Ye-Lun, Z., Guang-Yao, S., Hui-Juan, M., 2013. Effect of consumption of micronutrient enriched wheat steamed bread on postprandial plasma glucose in healthy and type 2 diabetic subjects. Nutr. J. 12, 64–72. www.biomedcentral.com/content/pdf/1475-2891-12-64.pdf (accessed 25.06.15.).

Tsai, S.Y., Yang, J.H., Tseng, Y.H., Lee, C.E., Mau, J.L., 2010. Quality of silver ear steamed bun. J. Food Process. Preservation 34, 649–663. http://dx.doi.org/10.1111/j.1745-4549.2009.00383.x.

Tseng, Y.H., Yang, J.H., Lee, C.E., Mau, J.L., 2011. Quality of shiitake stipe steamed bun. Czech J. Food Sci. 29, 79–86. http://agriculturejournals.cz/publicFiles/35020.pdf (accessed 28.09.15.).

van den Wijngaart, A., Codling, K., 2013. Impact of flour fortification on common Asian wheat flour-based foods. Cereal Food World 58, 238–245. http://dx.doi.org/10.1094/CFW-58-5-0238.

Xu, Y., Wang, L., He, J., Bi, Y., Li, M., Wang, T., Wang, L., Jiang, Y., Dai, M., Lu, J., Xu, M., Li, Y., Hu, M., Li, J., Mi, S., Chen, C.-S., Li, G., Mu, Y., Zhao, J., Ko, L., Chen, J., La, S., Wan, W., Zhao, W., Ning, G., 2013. Prevalence and control of diabetes in Chinese adults. J. Am. Med. Assoc. 310, 948–959.

Yang, Y.X., Wang, H.W., Cui, H.M., Wang, Y., Yu, L.D., Xiang, S.X., Zhou, S.Y., 2006. Glycemic index of cereals and tubers produced in China. World J. Gastroenterol 12, 3430–3433.

Yen, M.T., Tseng, Y.H., Lee, C.E., Mau, J.L., 2015. Quality of Enokitake supplemented steamed bun. Am. J. Adv. Food Sci. Technol. 3, 1–13. www.researchgate.net/profile/Jeng-Leun_Mau/publication/272396161_Quality_of_Enokitake_Supplemented_Steamed_Bun/links/54e5a08b0cf29865c33752ef.pdf (accessed 28.09.15.).

Chapter 13

Future Trends

13.1 CONSUMPTION PATTERNS

Diets in many Asian countries are becoming more westernized, with increased consumption of meat and dairy and an increased demand for more varied foods. These changes are due to increasing urbanization and transition to a market-based economy with higher incomes. Differences in consumer demand vary between rural and city population groups and different geographical regions, but there is already a much wider range of products available in supermarkets. This availability of both traditional and Western processed food reduces the need for time-consuming domestic meal preparation by busy families.

Wheat consumption has been steadily increasing in Southeast Asian countries such as Malaysia, Thailand, Vietnam, and the Philippines where up to recent times rice has been the stable food. The opposite trend is observed in northern China, where it is expected that wheat consumption will decrease from 65 kg in 2011–13 to 59 kg by 2023 (https://apps.fas.usda.gov/psdonline/psdHome.aspx). Given that the consumption of instant noodles and preprepared dumplings are increasing, it is expected that consumption of steamed bread in northern China will decrease over the period. At the same time, there will be an increase in consumption of Western breads, pastries, and cakes, which are growing from a low base. Baked products are seen as trendy and are available in supermarkets, international franchised coffee shops, and bakery chains such as BreadTalk, a Singaporean franchise. Many of these products are adapted to cater specifically for Asian preferences of flavor and texture and are readily available for consumption at breakfast or as snack food.

13.2 FUNDAMENTAL RESEARCH

In light of recent manufacturing industry developments, further fundamental research is needed to understand the influence of processing conditions on end product quality, as well as the impact of ingredients and additives. More work is also needed on end product quality evaluation and the degree to which objective methods can complement or replace sensory testing. This is particularly important because of the difficulties in training panels and maintaining proficiency and accuracy both within and between panels.

Steamed Breads. http://dx.doi.org/10.1016/B978-0-08-100715-0.00013-6

13.3 MEETING THE NEEDS OF THE CONSUMER

Given the wide assortment and variety of products in the market place, consumer choice will be based on appearance, taste, health, convenience, and price. Increasingly these are branded products which differentiate purchases. It is up to the manufacturer and those further up the supply chain to meet the demand for the raw materials of the right quality for the specific end use. Mechanized steamed product production requires the use of flours fit for the purpose with good mixing and processing tolerance and consistency of flour quality from delivery to delivery. It is important that parameters like protein content, ash content, water absorption, and dough strength are within tight specification tolerances.

Manufacturers must implement quality management and food safety programs to deliver products that are consistently of high quality and safe to eat. If, for example, flour of a different quality for a new product, or if higher protein content is required, it is up to the flour miller to purchase raw materials which will satisfy the food manufacturer's requirements. Most wheat-exporting countries implement classification and grading standards, thus offering information to buyers on basic grain quality characteristics and quality profile. An area of concern is mycotoxin contamination of some wheats because deoxynivalenol (DON) produced by the fungi is not destroyed during cooking, although higher levels are removed with the bran during flour milling (Zhang and Wang, 2014). The maximum level of DON is now regulated in a number of countries including China.

No value chain is complete without growers, wheat standards, segregation, storage and handling systems, and transport. Food safety and quality programs need to operate at all stages in the chain to deliver options to the flour miller to produce flour of the required quality for the manufacturer.

13.4 INDUSTRIAL PRODUCTION

Another challenge that will need to be met is maintaining product integrity and quality due to the increasing size of production facilities. Scaling up from small factories to meet ever growing demand will require larger plant and equipment and hence will require more stringent quality standards for ingredients and processes. Controlling dough functionality on a large scale will require research on mixing, proofing, and steaming regimes.

13.5 DEVELOPMENT OF NEW PRODUCTS

New product development is important for the availability of a range of products in the market place and to give consumers more choice. There are many possibilities for the future and many more new products will become available. For example, addition of functional ingredients such as fiber, phytochemicals, and antioxidants from grains and wholegrain products are reported to have health benefits (see Chapter 12). Addition of grains and seeds to bread is increasing in the West, and is feasible for steamed breads, but attention needs to be paid to the addition of properly processed grain, for example, large grains may need to be kibbled, cracked, or soaked. Inclusion of fiber, grains, and seeds to flour-based products often requires flour of higher protein content to "carry" the blended material. There are other opportunities to reduce the glycemic index and added sugar, thus helping to reduce the potential for obesity and type 2 diabetes (Chapter 12). Other changes to formulations could limit the use of chemical additives and promote clean labels which are being driven by the retail sector and consumer demands. Clean labels send a signal to consumers via packaging information that the product contains familiar and trusted simple and natural ingredients.

Another area of potential development is in sourdough fermentation. Sourdoughs have good flavor and aroma, and the long fermentation may lead to increased nutritional quality retarding starch digestibility and making minerals and phytochemicals more bioavailable (Poutanen et al., 2009).

13.6 IMPROVING SHELF LIFE

Attention has been drawn to the short shelf life of steamed bread several times in this book. Firmer texture and loss of flavor are the physical manifestations of staling and product spoilage results due to growth of microorganisms such as yeasts and molds. The trend to centralized manufacture and distribution has increased the need for longer shelf life. Experience with Western bread has shown that a number of strategies exist to extend shelf life. Improvement in flour quality and consistency of quality, use of preservatives and enzyme applications (preferably enabling clean labels), improvements in machinery, optimized processing conditions, attention to food hygiene during and after manufacture, packaging and storage solutions are some of the technologies that can be applied commercially. Most importantly, in evaluation of these technologies, the goal should be to produce products which have retained authentic quality characteristics. It should be possible to increase the shelf life of steamed bread to 4–5 days with careful management.

Frozen finished product technology has already extended the shelf life of high-value steamed products and a new market has developed in Asia to provide products for local consumption and to export these products to the West.

13.7 FROZEN DOUGH TECHNOLOGY

Frozen doughs and frozen finished end products are new developments in the manufacture of steamed bread. More research is needed into the flour quality characteristics required for frozen doughs, particularly with regard to protein content and protein quality. Guidelines are required for the management of the cold chain during storage, handling, transport, and retail operations to ensure the products reach the consumer in good condition.

13.8 GLUTEN-FREE PRODUCTS

There is an increasing requirement in the West for gluten-free products. This is due to the incidence of celiac disease and also the increasing numbers of gluten-intolerant or gluten-sensitive individuals. Celiac disease is an autoimmune disease triggered by the ingestion of gluten and is typically found in 1% of Western populations. It has been considered that celiac disease is rare in China, but research suggests it is potentially as high as 3.5% (Yuan et al., 2013) with many cases undiagnosed. Patients with celiac disease must follow a gluten-free diet to minimize symptoms and risks of secondary autoimmune diseases. Gluten-free products can contain no wheat, rye, or barley and but may contain, for example, maize, rice, sorghum, millet, and tapioca flour, or starch, together with hydrocolloids, oils, and sugar. The challenge is to produce a product that resembles the structure of steamed bread, but in the absence of gluten. This is an area of untapped research.

13.9 WHEAT BREEDING

In addition to yield and disease resistance, grain and flour quality and suitability for end use are important parameters for new varieties developed in wheat breeding programs. Traditionally, these parameters were measured (phenotyped) during the breeding program and only those with the required qualities progressed to the next stage. Biochemical and molecular advances in the last three decades have led to the development of markers for many characteristics. Once markers are identified, they can be used as a diagnostic tool to identify lines carrying the required markers. Use of marker-assisted selection reduces the amount of laboratory testing that is required in the early stages of breeding. A number of functional markers have been developed that are important for steamed bread quality, including various dough properties, flour color, product shape, and color (Xia et al., 2013). This area will continue to grow into the future as new technologies being developed will reduce the throughput time and cost of DNA assays, thus assisting in the selection of varieties with better and more targeted quality. Phenotypic evaluation is still required, utilizing rapid quality methods in the early stages of selection, but full quality evaluation, including end product testing, in the later stages of the breeding program.

In the future, this could see the development of genetically modified (GM) wheat with improved drought resistance or special health properties, subject to acceptability of GM wheat in different markets.

REFERENCES

Poutanen, K., Flander, L., Katina, K., 2009. Sourdough and cereal fermentation in a nutritional perspective. Food Microbiol. 26, 693–699.

Xia, X.C., He, Z.H., Jin, L.H., Wang, L.H., Zhang, Y., Li, G.Y., 2013. Understanding the effects of HMW-GS and LMW-GS on processing quality using near-isogenic lines and development of functional markers for LMW-GS in wheat. In: He, Z.H., Wang, D. (Eds.), Proceedings 11th International Gluten Workshop China, August 12–15, 2012. International Maize and Wheat Improvement Center (CIMMYT), Mexico, DF, pp. 107–112.

Yuan, J., Gao, J., Li, X., Liu, F., Wijmenga, C., Chen, H., Gilissen, L.J., 2013. The tip of the "celiac iceberg" in China: a systematic review and meta-analysis. Plos One 8 (12), e81151. http://dx.doi.org/10.1371/journal.pone.0081151.

Zhang, H., Wang, B., 2014. Fate of deoxynivalenol and deoxynivalenol-3-glucoside during wheat milling and Chinese steamed bread processing. Food Control 44, 86–91.

Index

'*Note*: Page numbers followed by "f" indicate figures and "t" indicate tables.'

Kolibri 360 GmbH
Ferdinand-Jühlke-Straße 7
99095 Erfurt, Deutschland
produktsicherheit@kolibri360.de